BEGINNING GAMES

Tea For Two

Number Games for Numerals, Number Sets, and Number Words

by Marilynn G. Barr

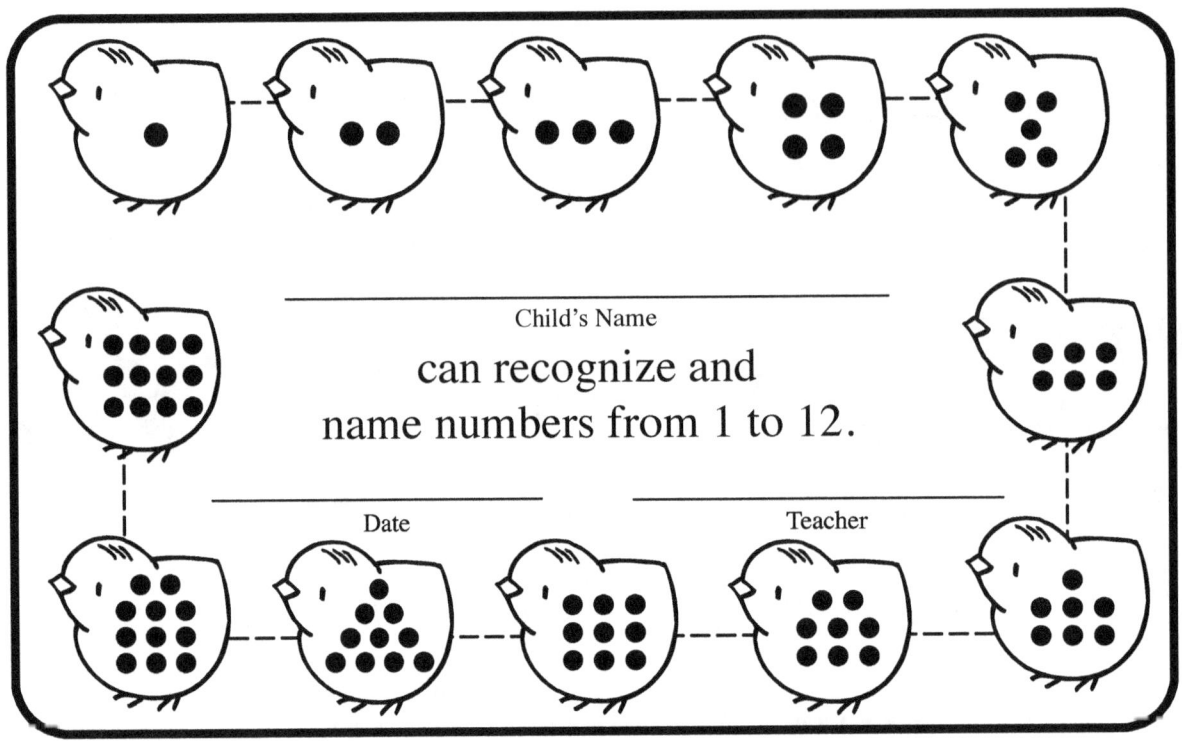

LAB201311
Beginning Games
TEA FOR TWO
by Marilynn G. Barr

Published by: Little Acorn Books™
Originally published by: Monday Morning Books, Inc.

Entire contents copyright © 2014 Little Acorn Books™

Little Acorn Books
PO Box 8787
Greensboro, NC 27419-0787

Promoting Early Skills for a Lifetime™

Little Acorn Books™
is an imprint of Little Acorn Associates, Inc.

http://www.littleacornbooks.com

Permission is hereby granted to reproduce student materials in this book for non-commercial individual or classroom use. *School-wide or system-wide use is expressly prohibited.

ISBN 978-1-937257-49-1

Printed in the United States of America

Contents

Introduction 4	Tiger Truck Tires 34
Bones For Bongo 5	Game Board 35
Game Board 6	Tiger Truck Patterns 36
Game Cards 8	Game Cards 37
Cover 10	Toast and Jelly 40
Turtle Races 11	Cover 41
Game Board 12	Game Board 42
Game Cards 14	Game Cards 44
Cover 16	Polar Bear Patches 46
Tea For Two 17	Cover 47
Game Board 18	Game Board 48
Game Cards 20	Game Cards 50
Cover 22	Cheep Cheep 53
Boots For Bunny 23	Game Board 54
Game Board 24	Game Cards 56
Bunny Patterns 25	Cover 58
Game Cards 26	Number Box Stackers 59
Birds of a Feather 29	Game Board 60
Game Board 30	Game Cards 62
Feather Patterns and Game Cards... 31	Cover 64

Tea For Two

Introduction

Children receive plenty of practice matching numerals, number words, and number sets with the ready-to-use beginning games featured in *Tea For Two*. Players learn to follow directions, practice fair-play, and develop readiness, fine-motor, and memory skills. Game formats include trail games, match boards, clothespin games, and stackers. Every game includes a game board and programmed playing pieces.

Children match numerals and number sets on trail games such as Bones for Bongo, Turtles Races, and Tea For Two. Boots For Bunny, Birds of a Feather, and Tiger Truck Tires clothespin games offer number recognition skills practice as well as fine motor skills development. Children clip matching clothespin game cards to bunnies, a peacock, and tiger trucks. Toast and Jelly, Polar Bear Patches, and Cheep Cheep match board games provide numeral and number word recognition. Players match jelly, patches, and chicks cutout game cards to the correct spaces on each game board. Number Box Stackers offers self-checking multi-dimensional skills practice as children identify and stack matching box game cards.

Tea For Two Tic-Tac-Toe For Two Players

Reproduce, color, and cut apart the game board and cards. Each player chooses the ladybug or leaf cards. In turn, each player places a card on one of the spaces. The first player with three ladybugs or leaves in a row, vertically, horizontally, or diagonally, wins.

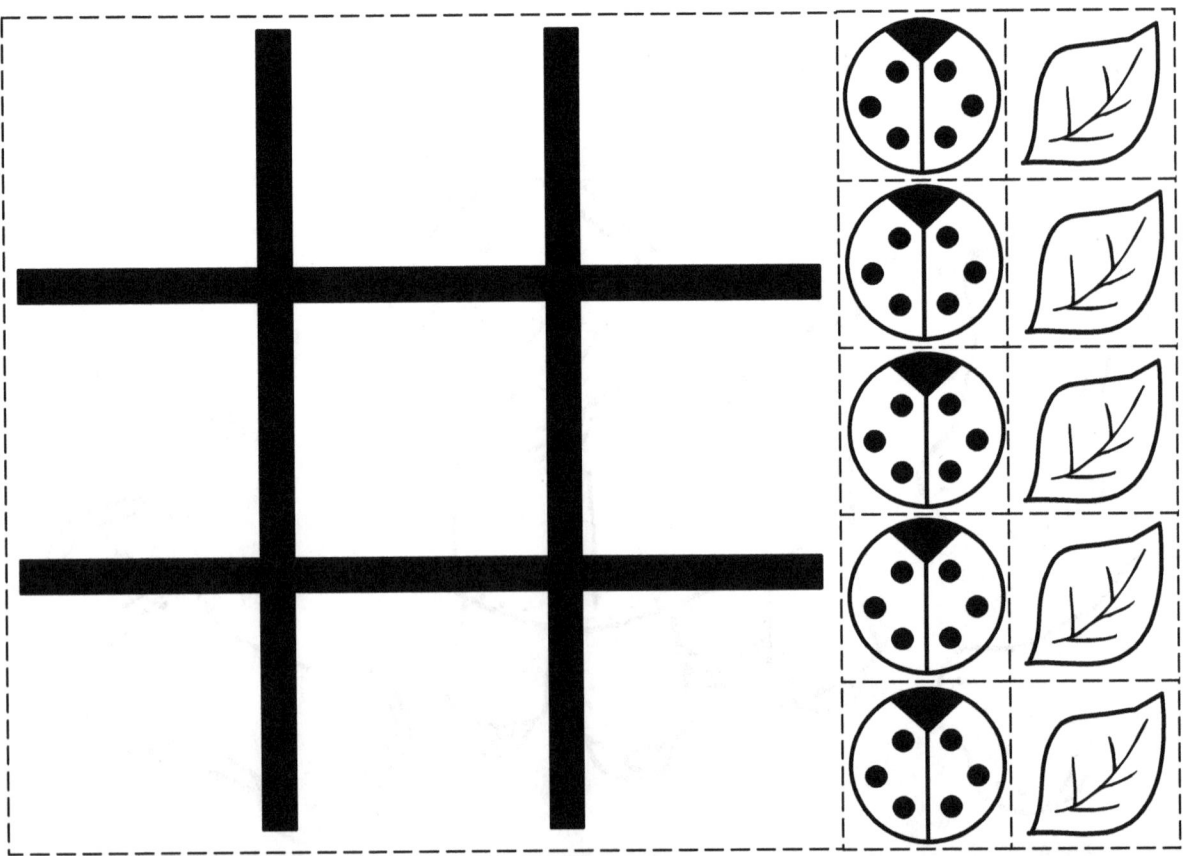

Bones For Bongo
A Trail Game
For Two to Four Players

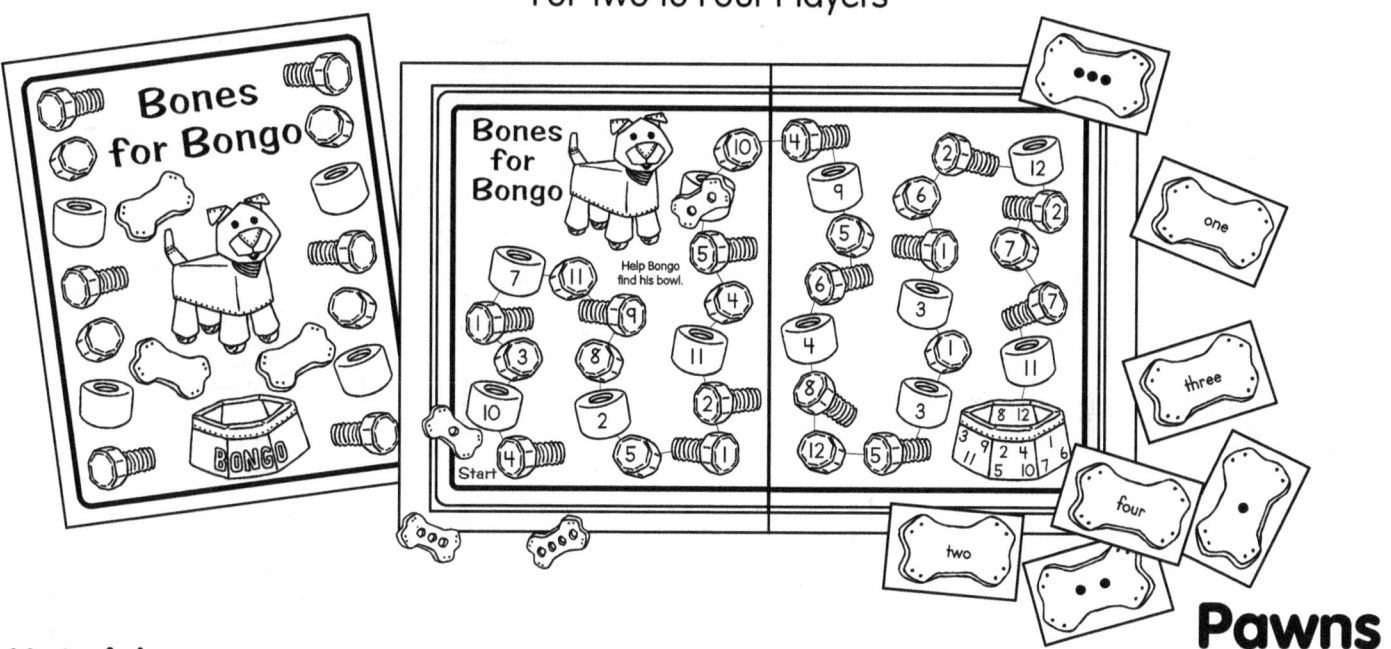

Pawns

Materials
crayons, markers, scissors, glue, file folder, envelope, tape

Assembly
Game Board: Reproduce, color, and cut out the cover and game board patterns. Matching in the center, glue the game board patterns to the inside of a folder. Glue the cover to the front of the folder, then laminate. Tape an envelope to the back of the game board folder to store pawns and game cards.

Pawns: Reproduce, color, laminate, and cut out a set of pawns. Store the pawns in the envelope on the back of the folder.

Game Cards: Reproduce, color, laminate, then cut apart two sets of the number set or number word game cards. Option: Reproduce, color, and glue each page of cards to the back of a sheet of gift wrap, then laminate, and cut apart the cards. Store the game cards in the envelope on the back of the game board folder.

How to Play
Set up the game board and cards on a table. Each player chooses a pawn. Then one player shuffles and places the deck of cards, face down, on the table. Each player, in turn, draws a card and moves his or her pawn to the next matching space on the game board. Drawn cards are placed, face down, in a discard pile. Play continues until each player reaches the dog bowl at The End. When all the cards have been drawn, reshuffle the discard pile and continue playing.

Bones For Bongo Game Board

Bones for Bongo

Help Bongo find his bowl.

Start

Bones For Bongo Game Board

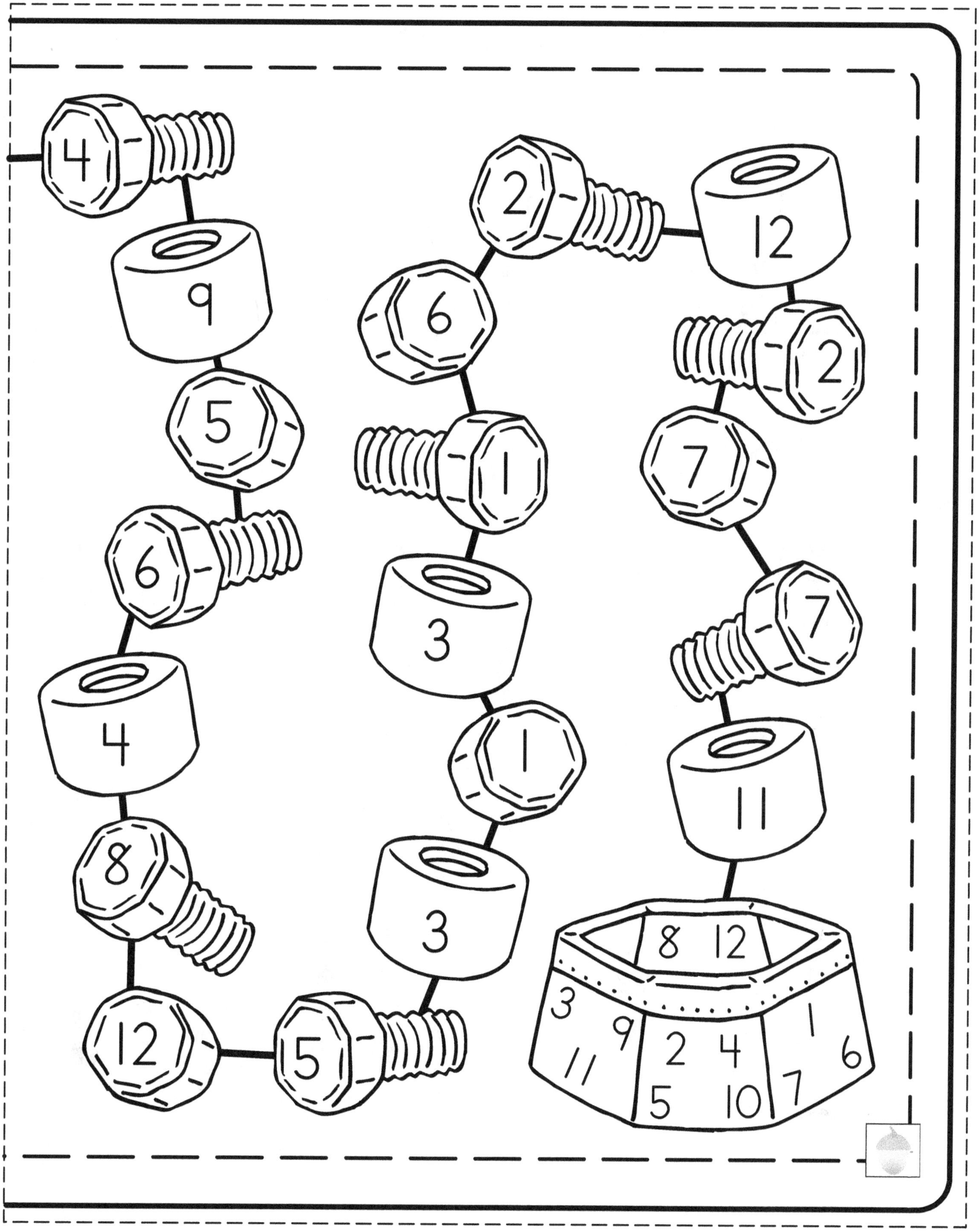

Number Set Bone Game Cards

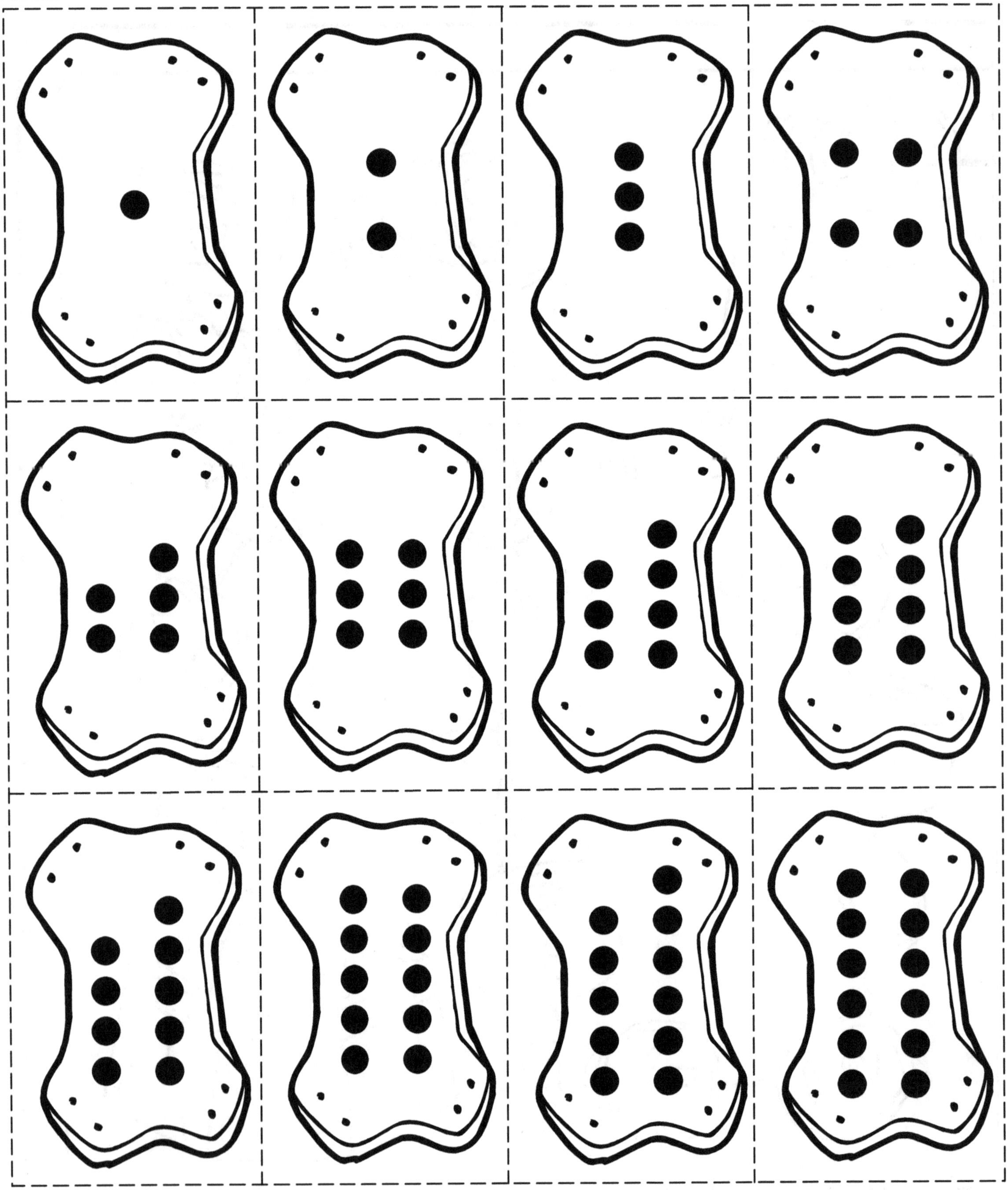

Reproduce two sets of game cards.

Number Word Bone Game Cards

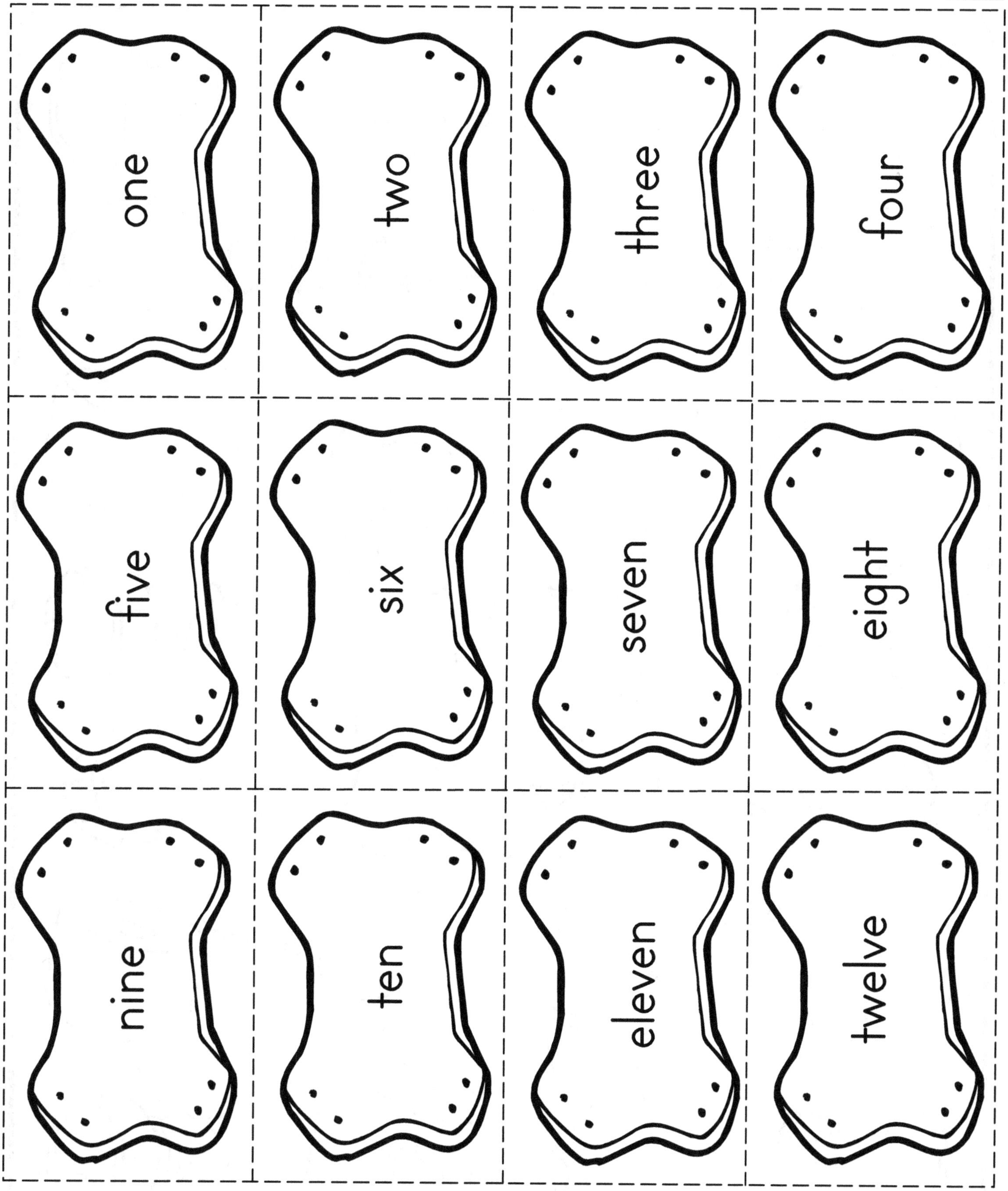

- one
- two
- three
- four
- five
- six
- seven
- eight
- nine
- ten
- eleven
- twelve

Reproduce two sets of game cards.

Bones for Bongo Cover

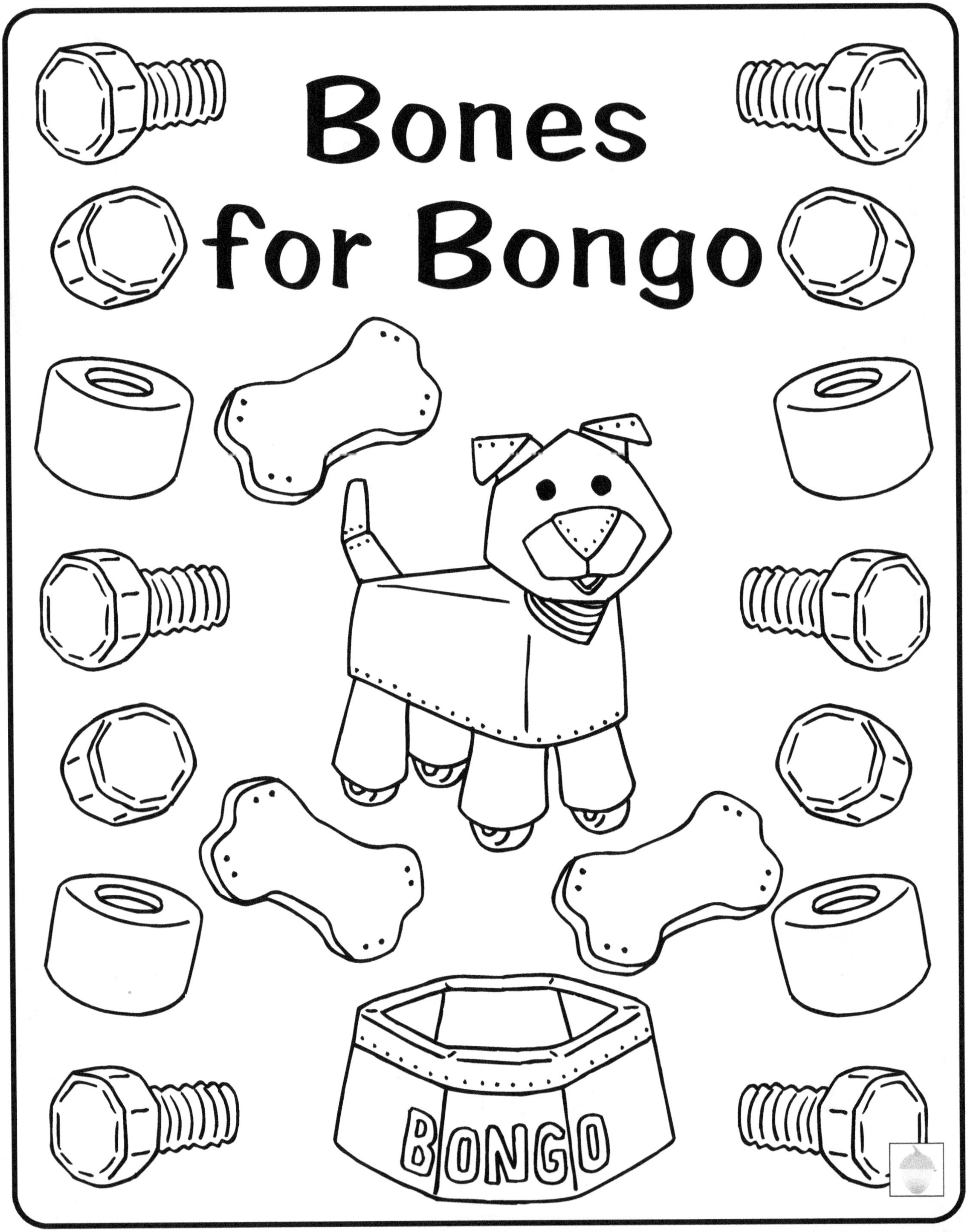

LAB201311 • TEA FOR TWO • 978-1-937257-49-1 •

Turtle Races
A Trail Game
For Two to Four Players

Pawns

Materials

crayons, markers, scissors, glue, file folder, envelope, tape

Assembly

Game Board: Reproduce, color, and cut out the cover and game board patterns. Matching in the center, glue the game board patterns to the inside of a folder. Glue the cover to the front of the folder, then laminate. Tape an envelope to the back of the game board folder to store pawns and cards.

Pawns: Reproduce, color, laminate, and cut out a set of pawns. Store the pawns in the envelope on the back of the folder.

Game Cards: Reproduce, color, laminate, then cut out two sets of number word or numeral game cards. Option: Reproduce, color, and glue each page of cards to the back of a sheet of gift wrap, then laminate and cut out the cards. Store the game cards in the envelope on the back of the game board folder. (Include a mix of number set and number word game cards for advanced players.)

How to Play

Set up the game board and cards on a table. Each player chooses a pawn. Then one player shuffles and places the game cards, face down, on the table. Each player, in turn, draws a card and moves his or her pawn to the next matching space on the game board. Drawn cards are placed, face down, in a discard pile. Play continues until each player reaches The End. When all the cards have been drawn, reshuffle the discard pile and continue playing.

Turtle Races Game Board

Turtle Races Game Board

Start

Turtle Game Cards

Reproduce, color, and cut out two sets of game cards.
Creative Options: Reproduce turtles for children to color and cut out.

Turtle Game Cards

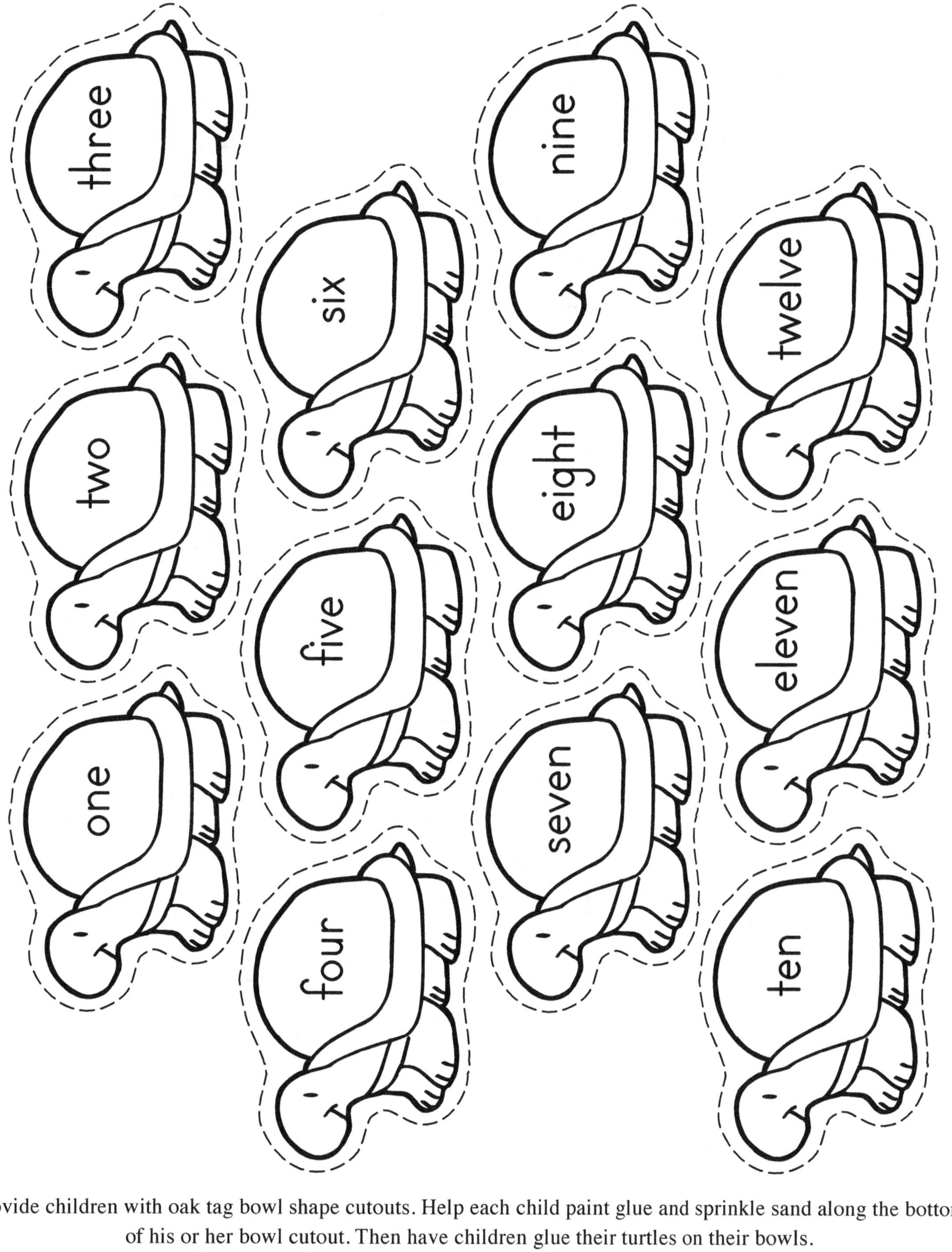

three, two, one, six, five, four, nine, eight, seven, twelve, eleven, ten

Provide children with oak tag bowl shape cutouts. Help each child paint glue and sprinkle sand along the bottom of his or her bowl cutout. Then have children glue cards turtles on their bowls.

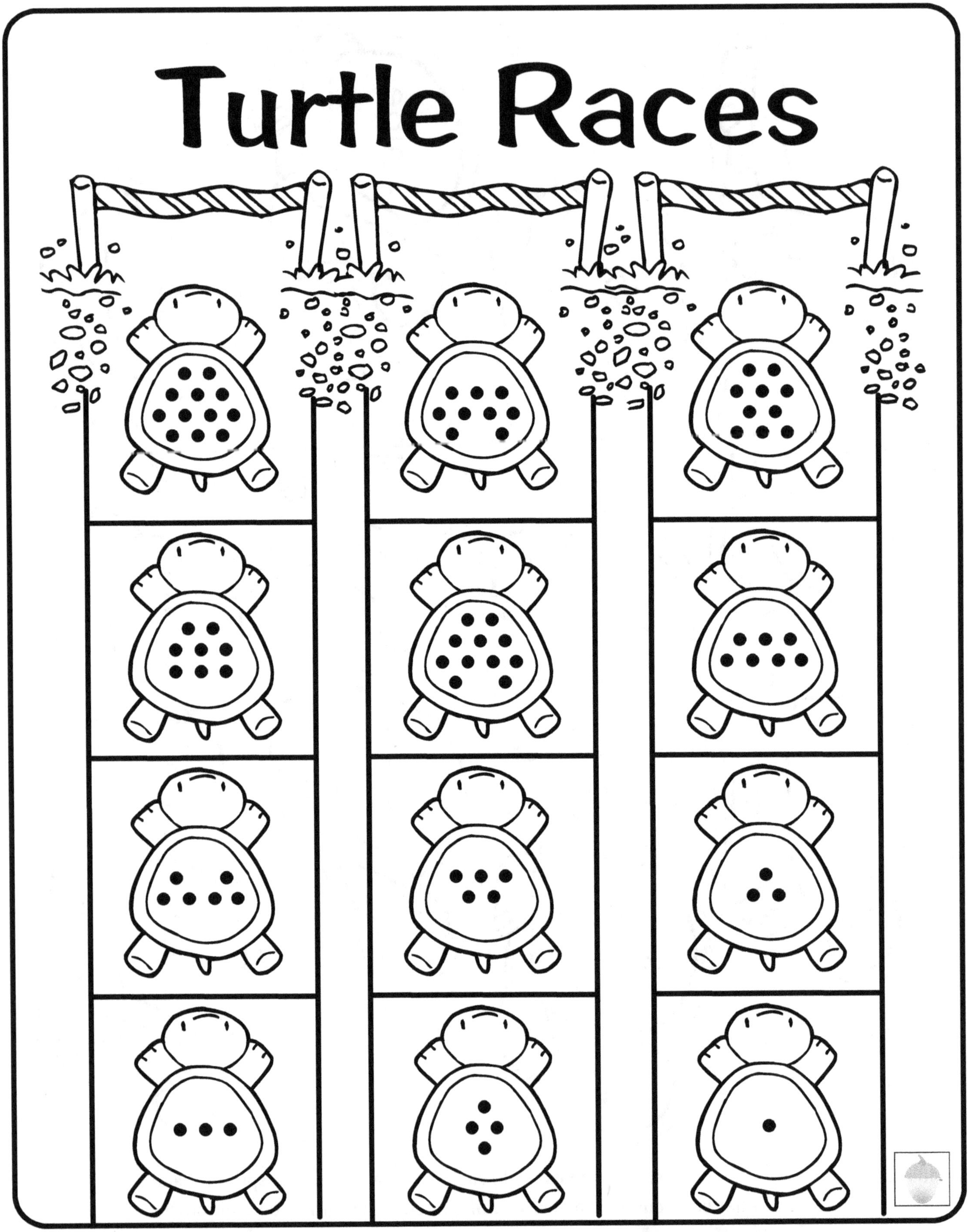

Turtle Races Cover

Turtle Races

Tea For Two
A Trail Game
For Two to Four Players

Pawns

Materials

crayons, markers, scissors, glue, file folder, envelope, tape

Assembly

Game Board: Reproduce, color, and cut out the cover and game board patterns. Matching in the center, glue the game board patterns to the inside of a folder. Glue the cover to the front of the folder, then laminate. Tape an envelope to the back of the game board folder to store pawns and game cards.

Pawns: Reproduce, color, laminate, and cut out a set of pawns. Store the pawns in the envelope on the back of the folder.

Game Cards: Reproduce, color, laminate, then cut out two sets of number set or number word ladybug game cards. Option: Reproduce, color, and glue each page of cards to the back of a sheet of gift wrap, then laminate and cut apart the cards. Store the game cards in the envelope on the back of the game board folder. (Include a mix of number word and number set game cards for advanced players.)

How to Play

Set up the game board and cards on a table. Each player chooses a pawn. Then one player shuffles and places the lady bug game cards, face down, on the table. Each player, in turn, draws a card and moves his or her pawn to the next matching space on the game board. Drawn cards are placed, face down, in a discard pile. Play continues until each player reaches the teapot at The End. When all the cards have been drawn, reshuffle the discard pile and continue playing.

Tea For Two Game Board

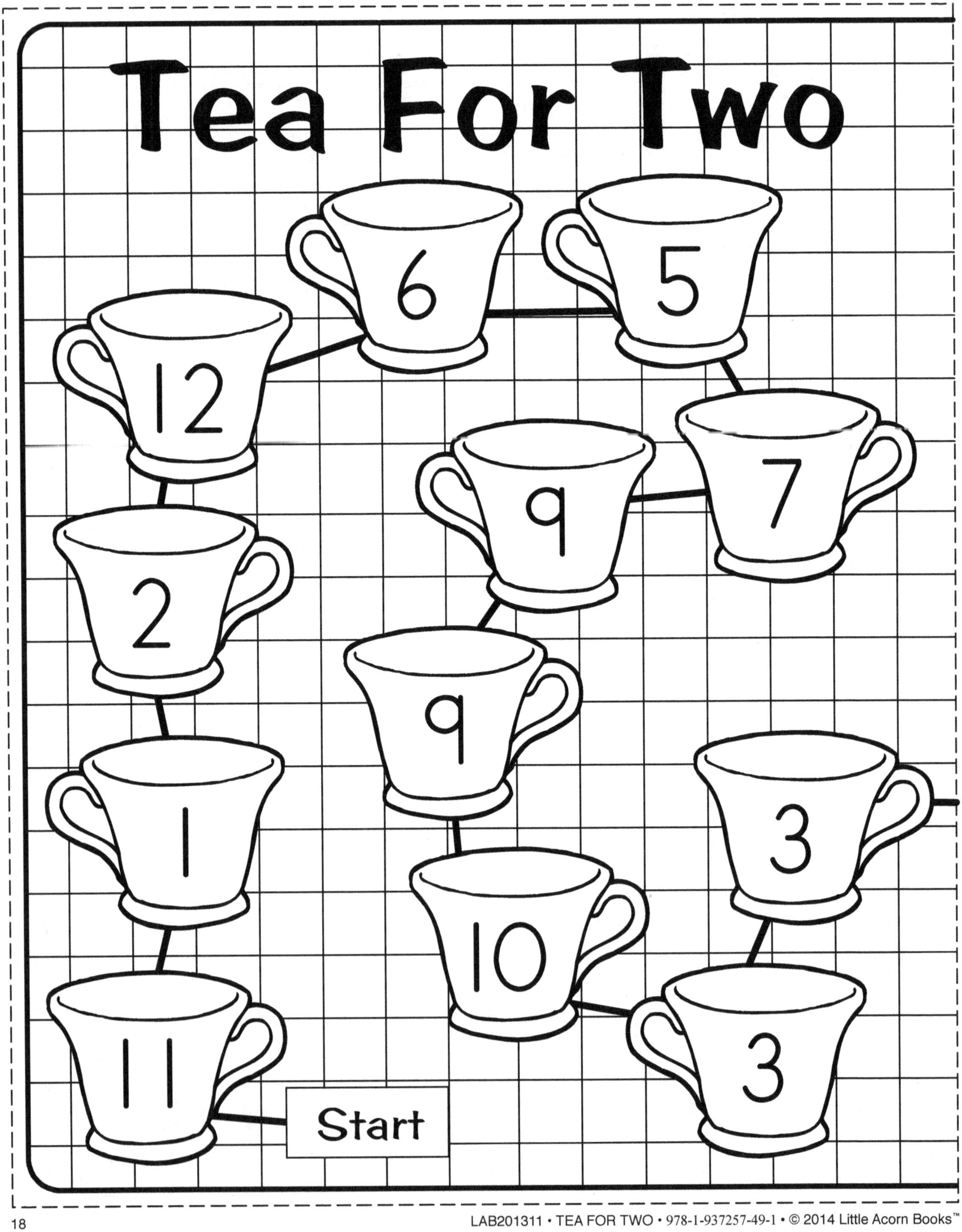

LAB201311 • TEA FOR TWO • 978-1-937257-49-1 • © 2014 Little Acorn Books™

Tea For Two Game Board

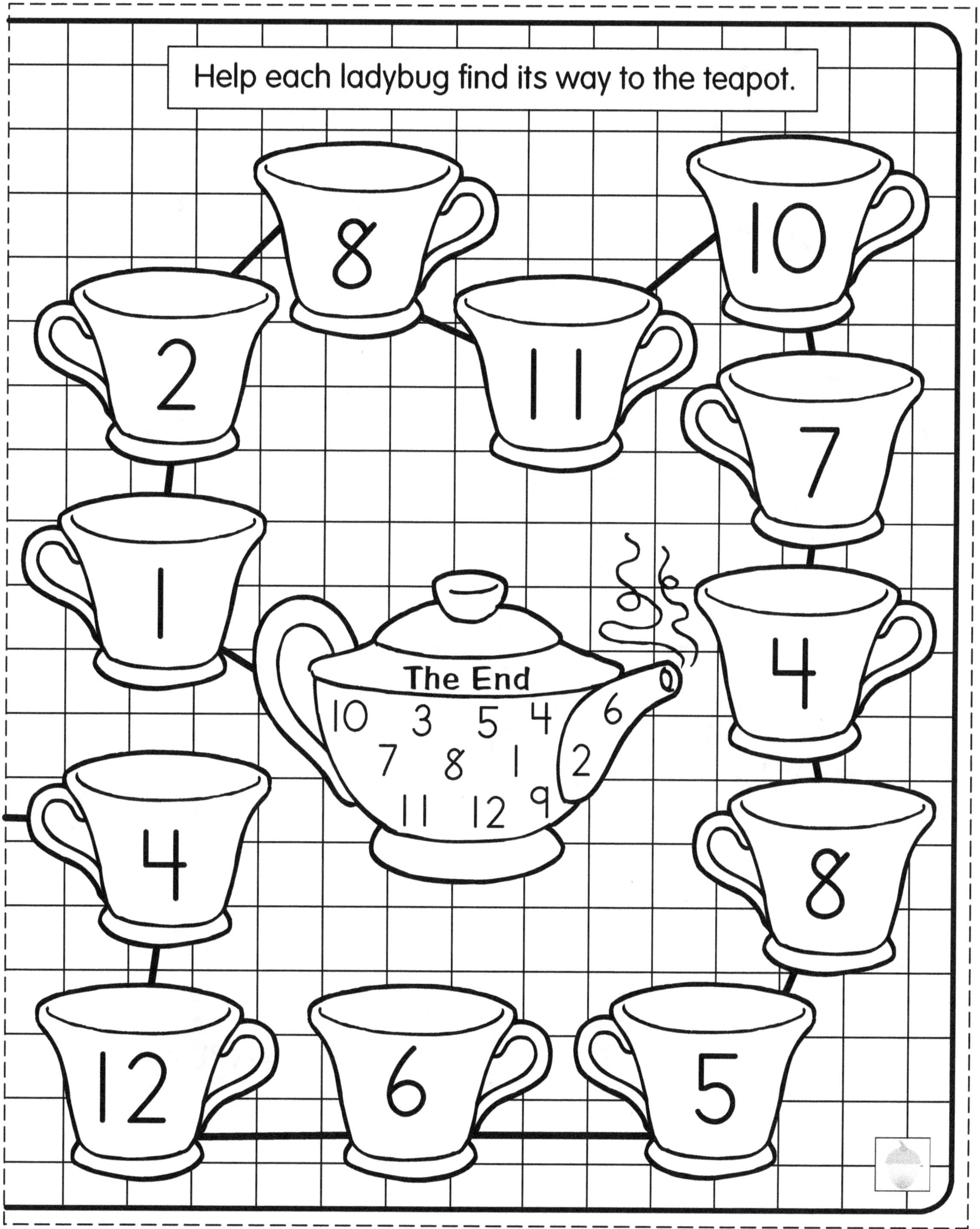
Help each ladybug find its way to the teapot.
8
10
2
11
7
1
4
The End
10 3 5 4 6
7 8 1 2
11 12 9
4
8
12
6
5

Ladybug Game Cards

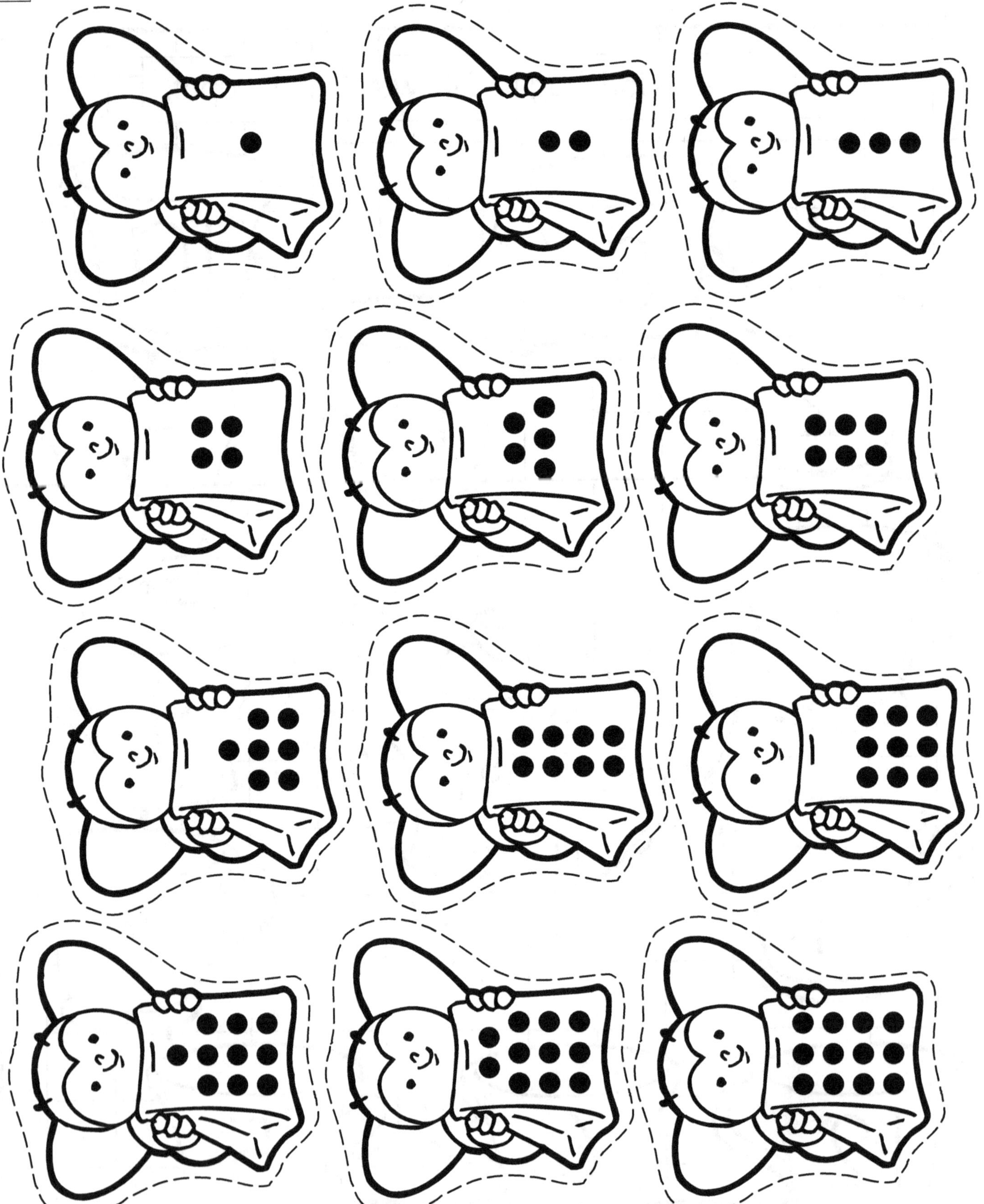

Reproduce, color, and cut out two sets of game cards.

Ladybug Game Cards

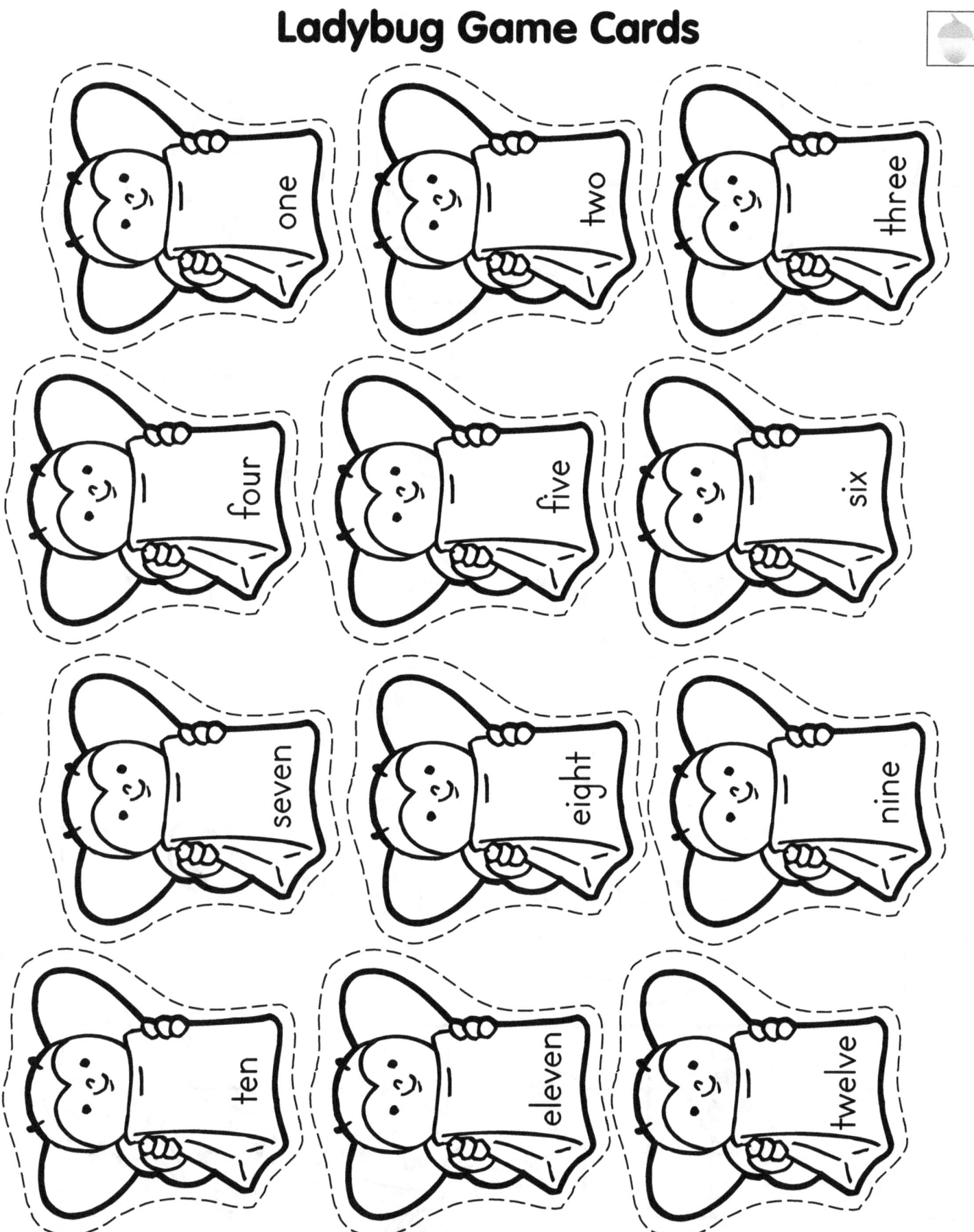

Creative Options: Make a counting chart to display in a skills practice center. Reproduce, color, cut out, and glue ladybug game cards, in sequence, on an oak tag sentence strip.

Tea For Two Cover

Boots For Bunny
A Clothespin Game
For Two Players

Materials
crayons, markers, scissors, glue, file folders, clothespins, large envelope

Assembly
Game Board: Reproduce, color, and cut out four game board patterns. Matching along the straight edges, glue the game board patterns on a poster board circle to form a round game board. Glue the title in the center of the game board. Reproduce, color, cut out, and glue eight bunny patterns around the game board. Program each bunny's tummy with a numeral. Make additional game boards to focus on different numbers.

Clothespin Game Cards: Reproduce, color, and cut out a set of numeral, number word, or number set boot game cards. Glue a clothespin to the back of each game card. Decorate a large envelope with bunny patterns and boots. Store the clothespin game cards in the envelope. (Include a mix of number word and number set game cards for advanced players.)

How to Play
Set up the game board on a table. Place the clothespin game cards, face down, on the table. Each player, in turn, draws a boot clothespin card. If there is a match, the player clips the matching clothespin boot to the correct bunny. If there is no match, the player places the boot back on the table, face down. Play continues until every bunny is wearing a pair of matching boots.

Boots For Bunny Game Board and Title

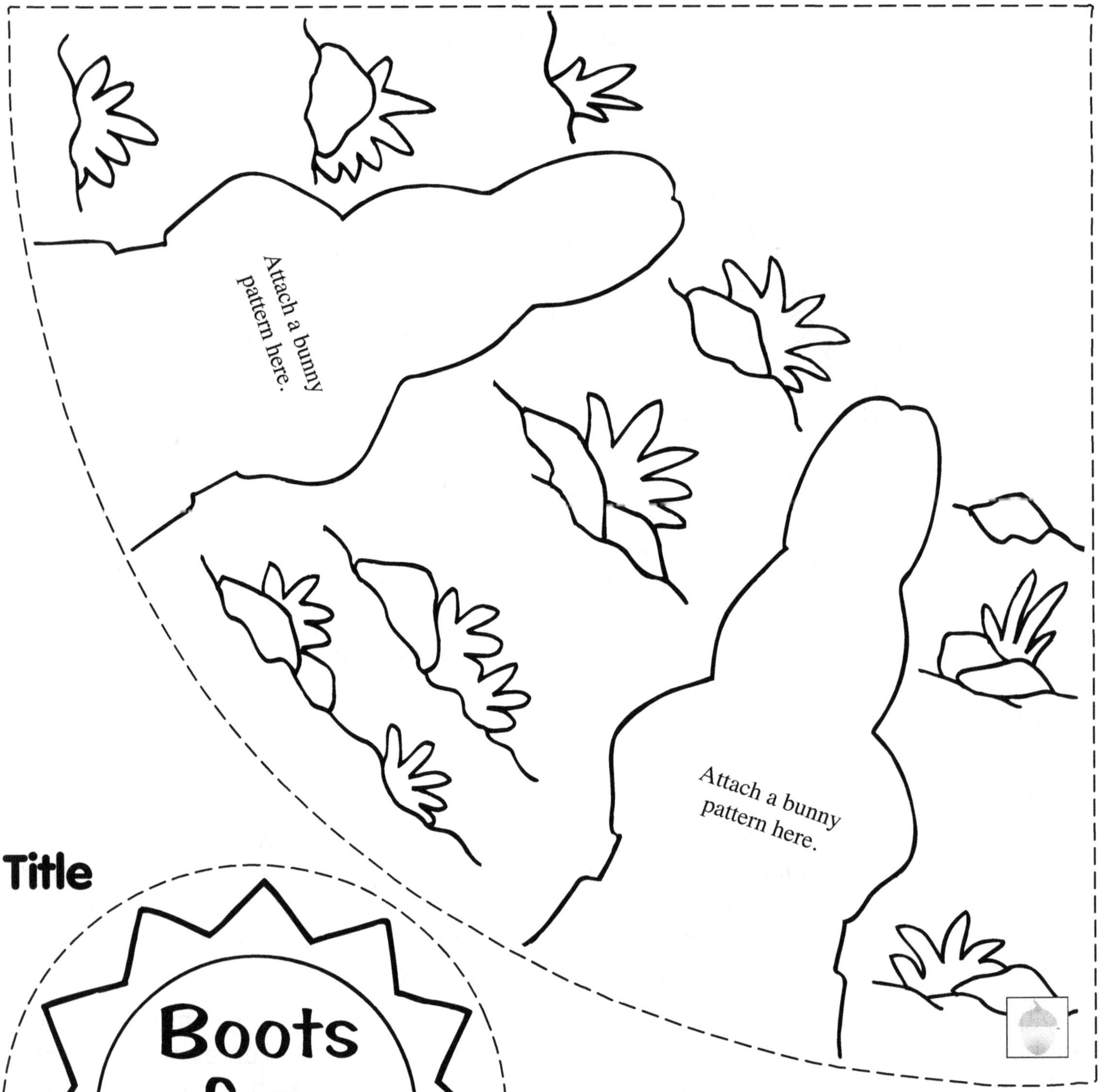

Attach a bunny pattern here.

Attach a bunny pattern here.

Title

Boots for Bunny

Reproduce, color, cut out, and assemble four game board patterns on a poster board circle to form a round game board. Glue the title in the center of the game board.

Bunny Patterns

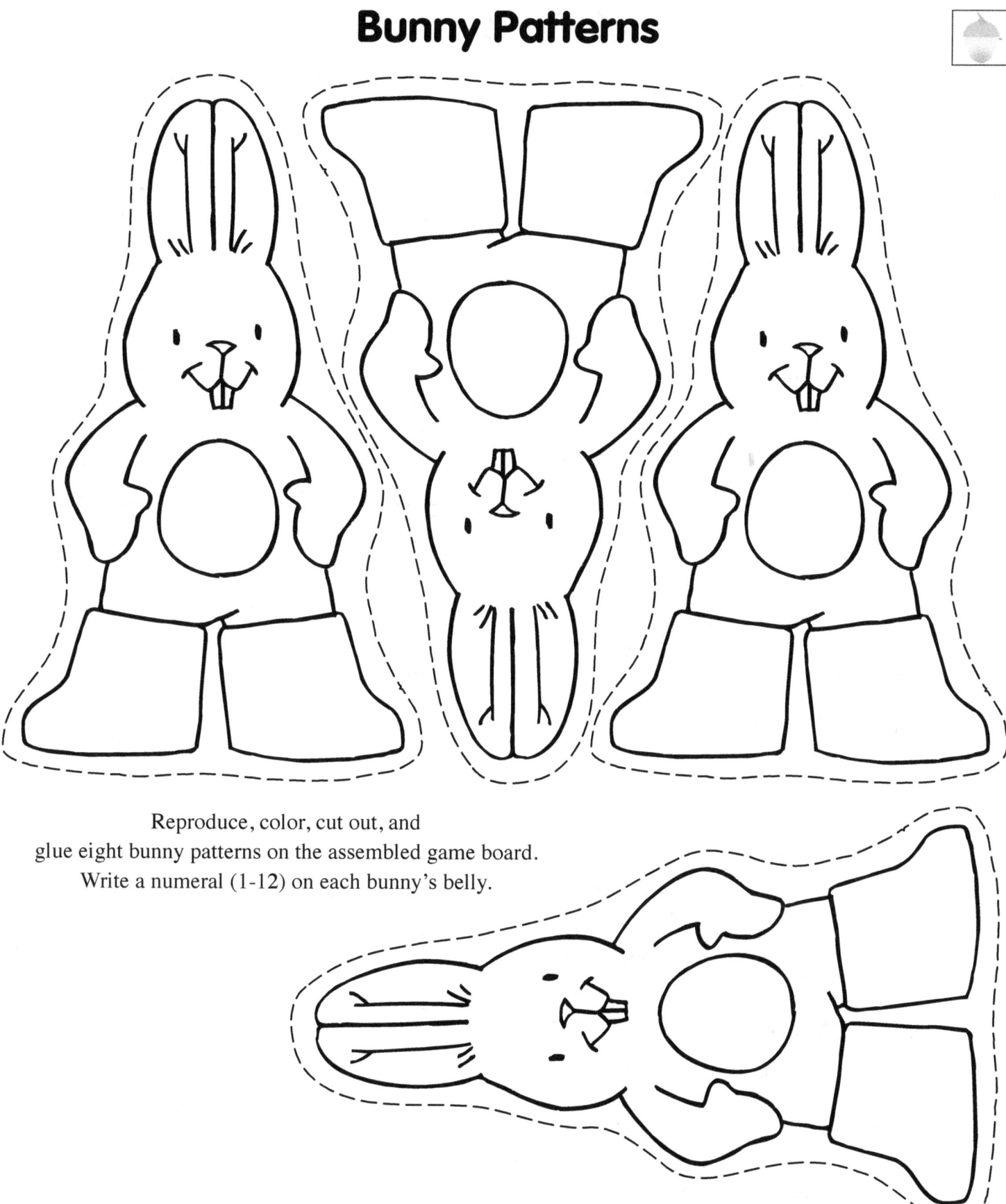

Reproduce, color, cut out, and
glue eight bunny patterns on the assembled game board.
Write a numeral (1-12) on each bunny's belly.

Creative Option: Reproduce, color, and cut out a bunny pattern.
Decorate the bunny's boots with glitter, buttons, yarn scraps, or pom poms.
Glue a cotton ball on the bunny's belly. Glue a craft stick to the back of the bunny to form a stick puppet.

Boot Game Cards

Reproduce, color, and cut out one set of cards.

Boot Game Cards

Reproduce, color, and cut out one set of cards.

Number Set Boot Game Cards

Reproduce, color, and cut out one set of cards.

Birds of a Feather
A Clothespin Game
For Two Players

Materials
crayons, markers, scissors, glue, file folders, clothespins, large envelope

Assembly
Game Board: Reproduce, color, and cut out the game board pattern. Glue the game board pattern on a sheet of poster board. Reproduce, color, cut out, and glue a set of numeral, number set, or number word feathers around the peacock. Trim away the excess poster board around the peacock and feathers. Make additional game boards to focus on different numbers. See the note on page 30.

Clothespin Game Cards: Reproduce, color, and cut out a set of numeral, number set, or number word teardrop game cards. Glue a clothespin to the back of each game card. Decorate a large envelope with feather game cards. Store the clothespin game cards in the envelope. (Include non-matching game cards for advanced players, or all matches for early learners.)

How to Play
Set up the game board on a table. Place the clothespin game cards, face down, on the table. Each player, in turn, draws a teardrop clothespin card. If there is a match, the player clips the matching card to the correct feather. If there is no match, the player places the card back on the table, face down. Play continues until a teardrop card is clipped to all of the peacock's feathers.

Peacock Game Board

Reproduce, color, cut out, and glue a set of numeral, number set, or number word feathers around the peacock. Then reproduce a set of numeral, number set, or number word tear-drop game cards.

Note: Children can match dots to dots, numerals to numerals, number words to number words, etc.

Birds of a Feather

Creative Option: Reproduce, color, cut out, and glue the peacock in the center of a sheet of construction paper. Glue craft tail feathers to the peacock. Glue a glitter border around the picture.

Feather Patterns and Teardrop Game Cards

Feather Patterns and Teardrop Game Cards

Feather Patterns and Teardrop Game Cards

one • two • three • four • five • six • seven • eight • nine • ten • eleven • twelve

Tiger Truck Tires
A Clothespin Game
For Two Players

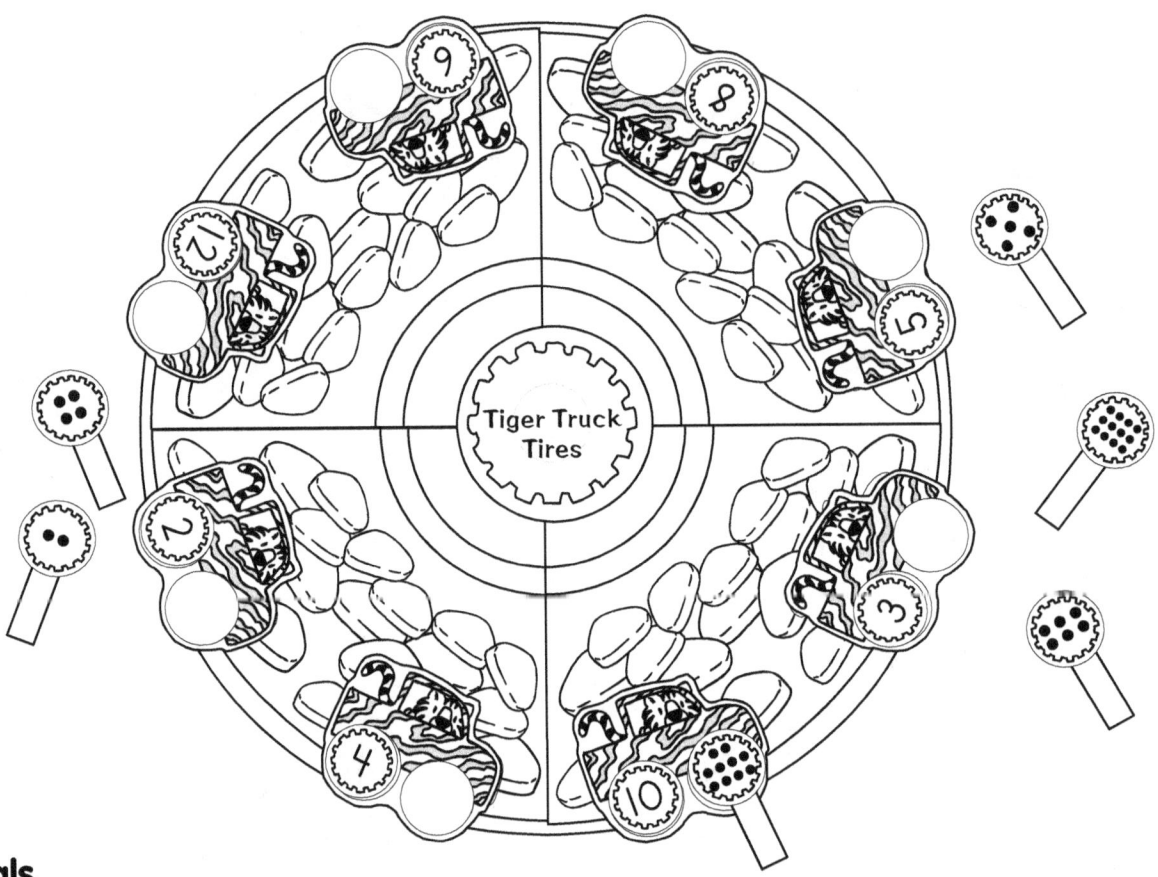

Materials
crayons, markers, scissors, glue, file folders, clothespins, large envelope

Assembly
Game Board: Reproduce, color, and cut out four game board patterns. Matching along the straight edges, glue the game board patterns on a poster board circle to form a round game board. Glue the title in the center of the game board. Reproduce, color, cut out, and glue eight tiger truck patterns around the game board. Glue one numeral, number word, or number set game card each truck. Make additional game boards to focus on different numbers.

Clothespin Game Cards: Reproduce, color, and cut out a set of numeral, number word, or number set tire game cards. Glue a clothespin to the back of each game card. Decorate a large envelope with tire game cards. Store the clothespin game cards in the envelope.

How to Play
Set up the game board on a table. Place the clothespin game cards, face down, on the table. Each player, in turn, draws a tire clothespin card. If there is a match, the player clips the matching clothespin tire to the correct truck. If there is no match, the player places the tire back on the table, face down. Play continues until every truck has a pair of matching tires.

Tiger Truck Tires Game Board

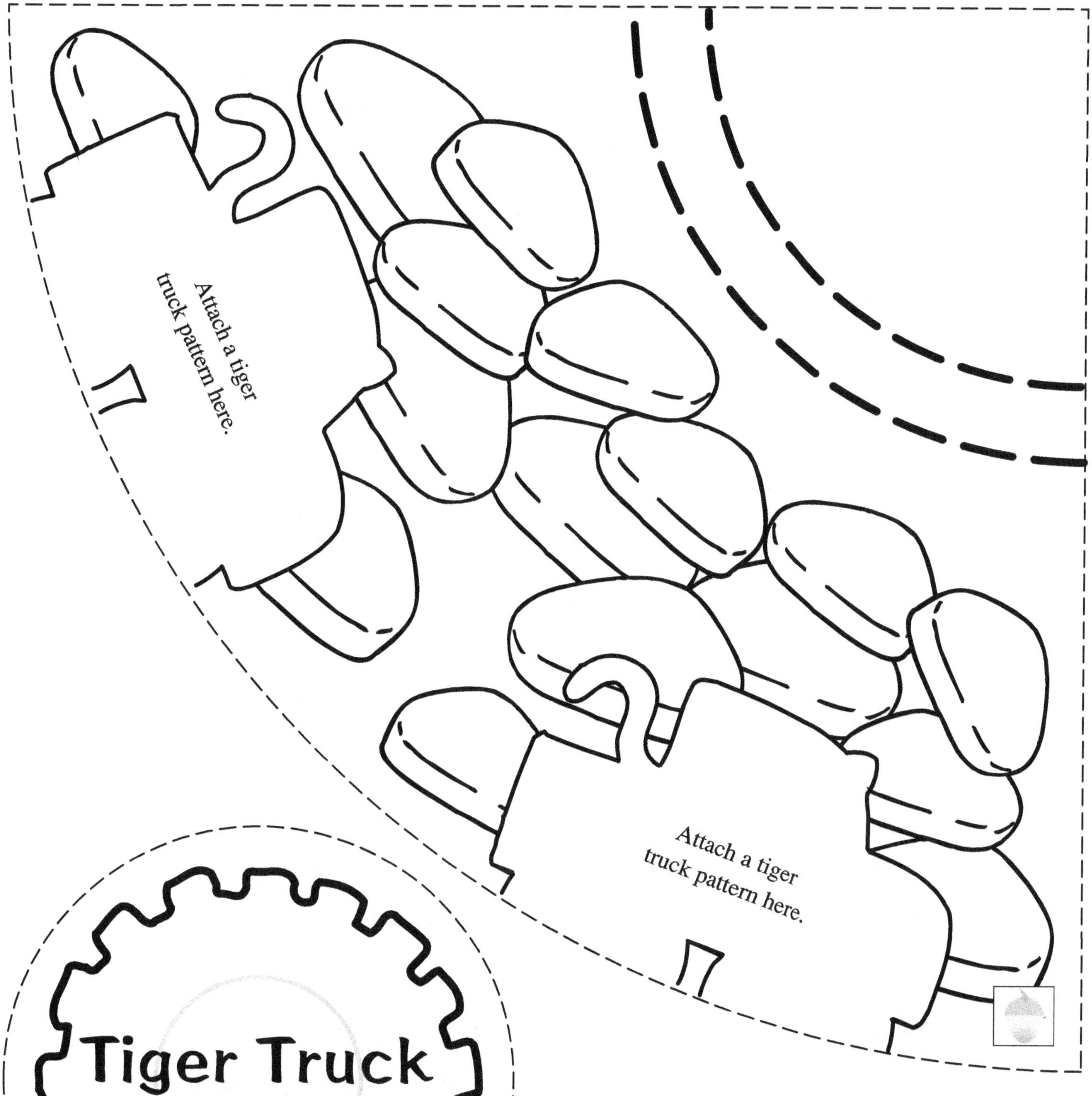

Attach a tiger truck pattern here.

Attach a tiger truck pattern here.

Tiger Truck Tires

Reproduce, color, cut out, and assemble four game board patterns on a poster board circle to form a round game board. Glue the title in the center of the game board.

Tiger Truck Patterns

Reproduce two sets of tiger truck patterns.

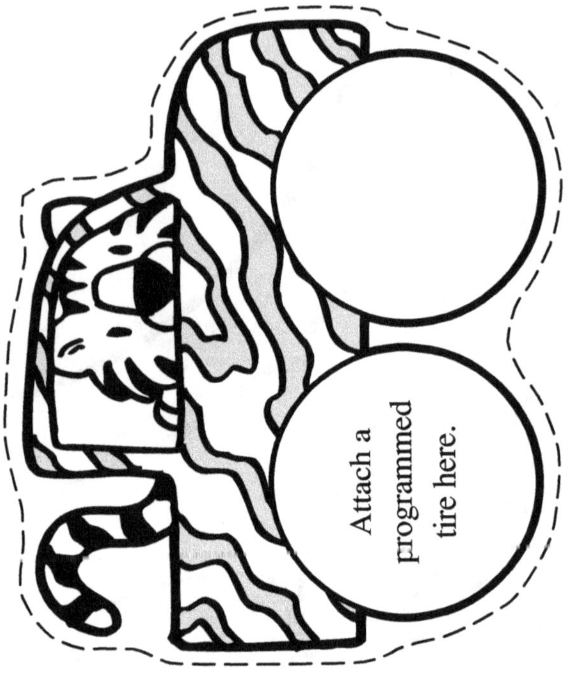

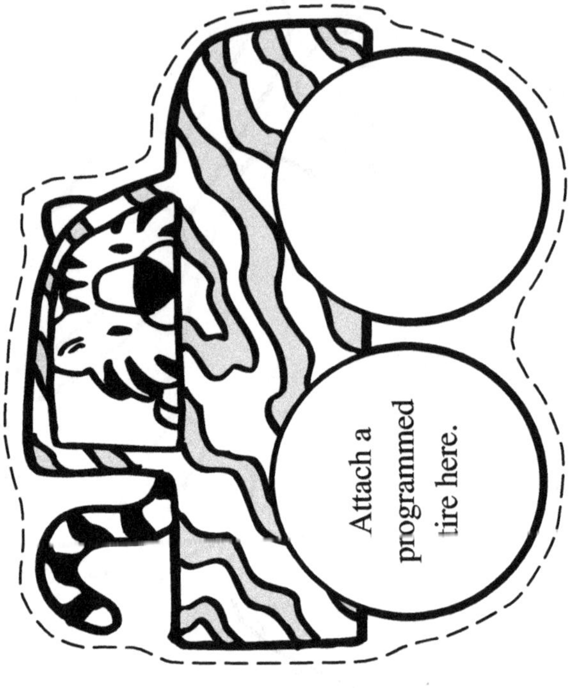

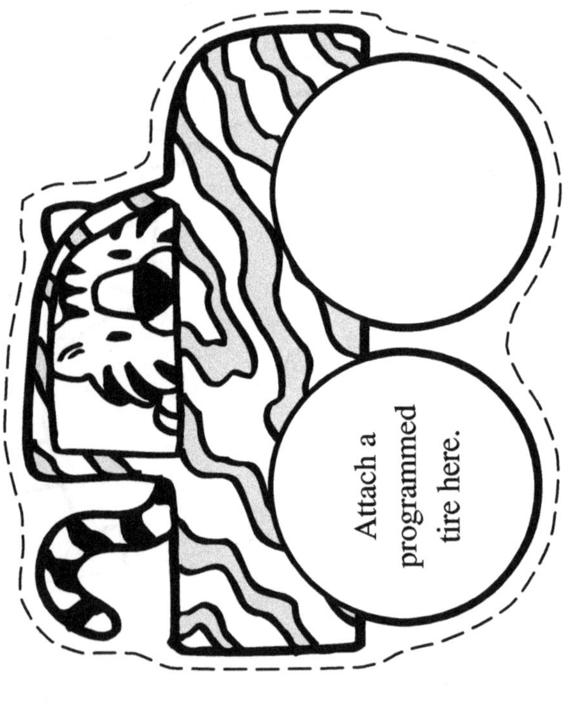

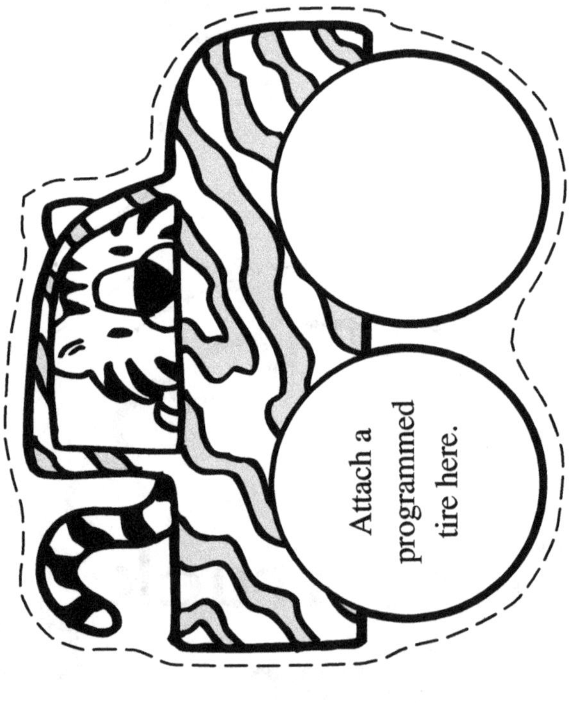

Attach a programmed tire here.

Attach a programmed tire here.

Attach a programmed tire here.

Attach a programmed tire here.

Creative Option: Provide children with a sheet of construction paper and a length of scalloped bulletin board border to form a bumpy road picture. Have children color, cut out, and assemble matching sets of tire game cards on tiger truck cutouts. Help each child glue the bulletin board border on a sheet of construction paper, then glue his or her trucks on the bumpy road.

Tire Game Cards

Reproduce one set of tire game cards.

Tire Game Cards

Reproduce one set of tire game cards.

1 2 3
4 5 6
7 8 9
10 11 12

38   LAB201311 • TEA FOR TWO • 978-1-937257-49-1 • © 2014 Little Acorn Books™

Tire Game Cards

Reproduce one set of tire game cards.

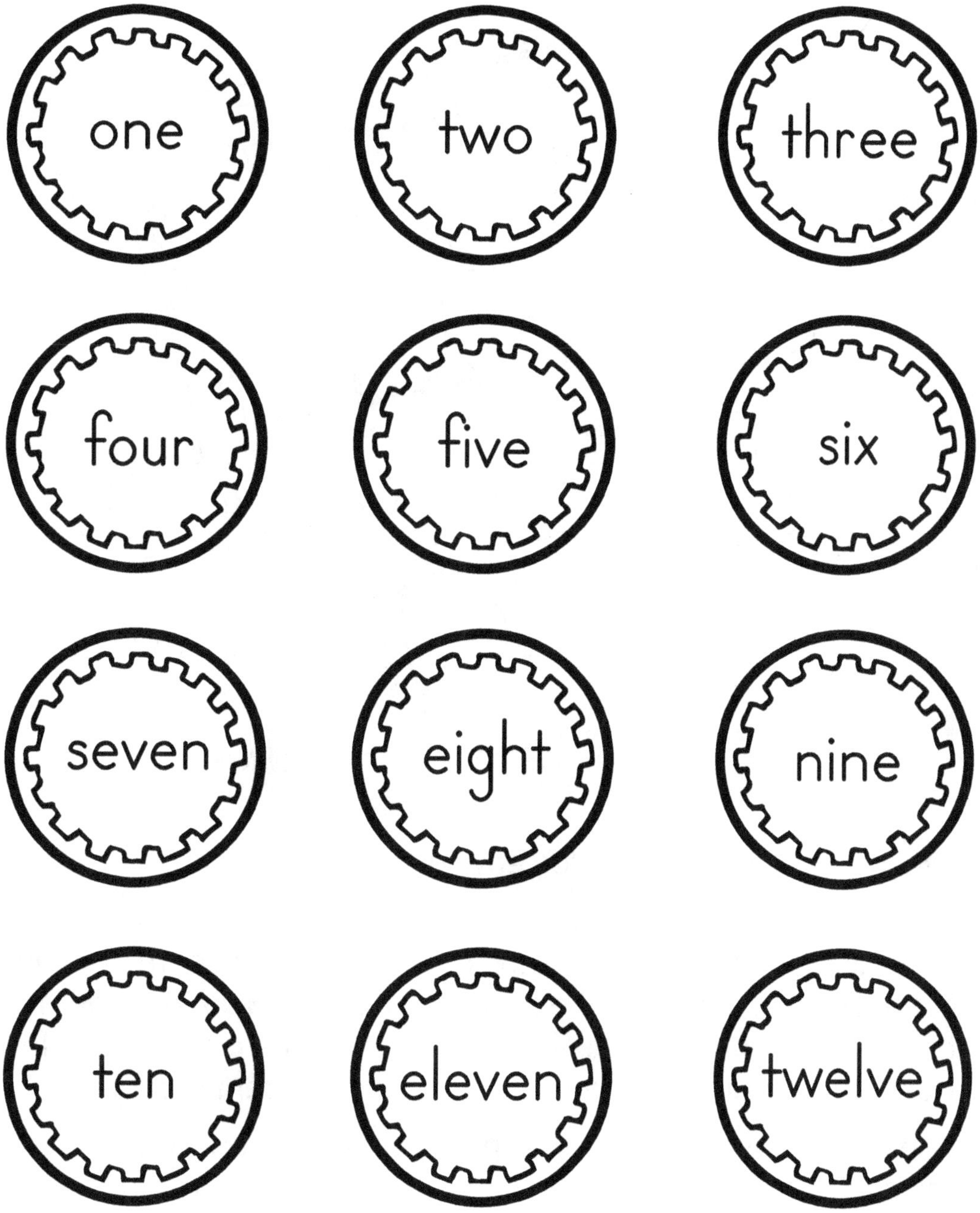

one	two	three
four	five	six
seven	eight	nine
ten	eleven	twelve

Toast and Jelly
A Match Board Game
For Two Players

Materials
crayons, markers, scissors, glue, file folder, envelope

Assembly
Game Board: Reproduce, color, and cut out the cover and game board patterns. Assemble and glue the game board patterns to the inside of a folder. Glue the cover to the front of the folder, then laminate. Tape an envelope to the back of the game board folder to store game cards. Make additional game boards to focus on different numbers.

Game Cards: Reproduce, color, laminate, then cut out a set of numeral or number set jelly game cards. Option: Reproduce, color, and glue a page of game cards to the back of a sheet of gift wrap, then laminate and cut out the cards. Store the game cards in the envelope on the back of the game board folder. (Include a mix of numeral and number set jelly game cards for advanced players.)

How to Play
Set up the game board on a table. One player shuffles and places the jelly game cards, face down, on the table. Each player, in turn, draws a card. If there is a match, the player identifies the match, and places the card on the correct slice of toast. If there is no match, the player places the card, face down, in a discard pile. Play continues until a matching jelly game card has been placed on every slice of toast. Reshuffle the discarded jelly game cards if needed.

Toast and Jelly Cover

Toast and Jelly Game Board

Toast and Jelly Game Board

six

five

eight

three

one

twelve

Spread jelly on each slice of toast.

Jelly Game Cards

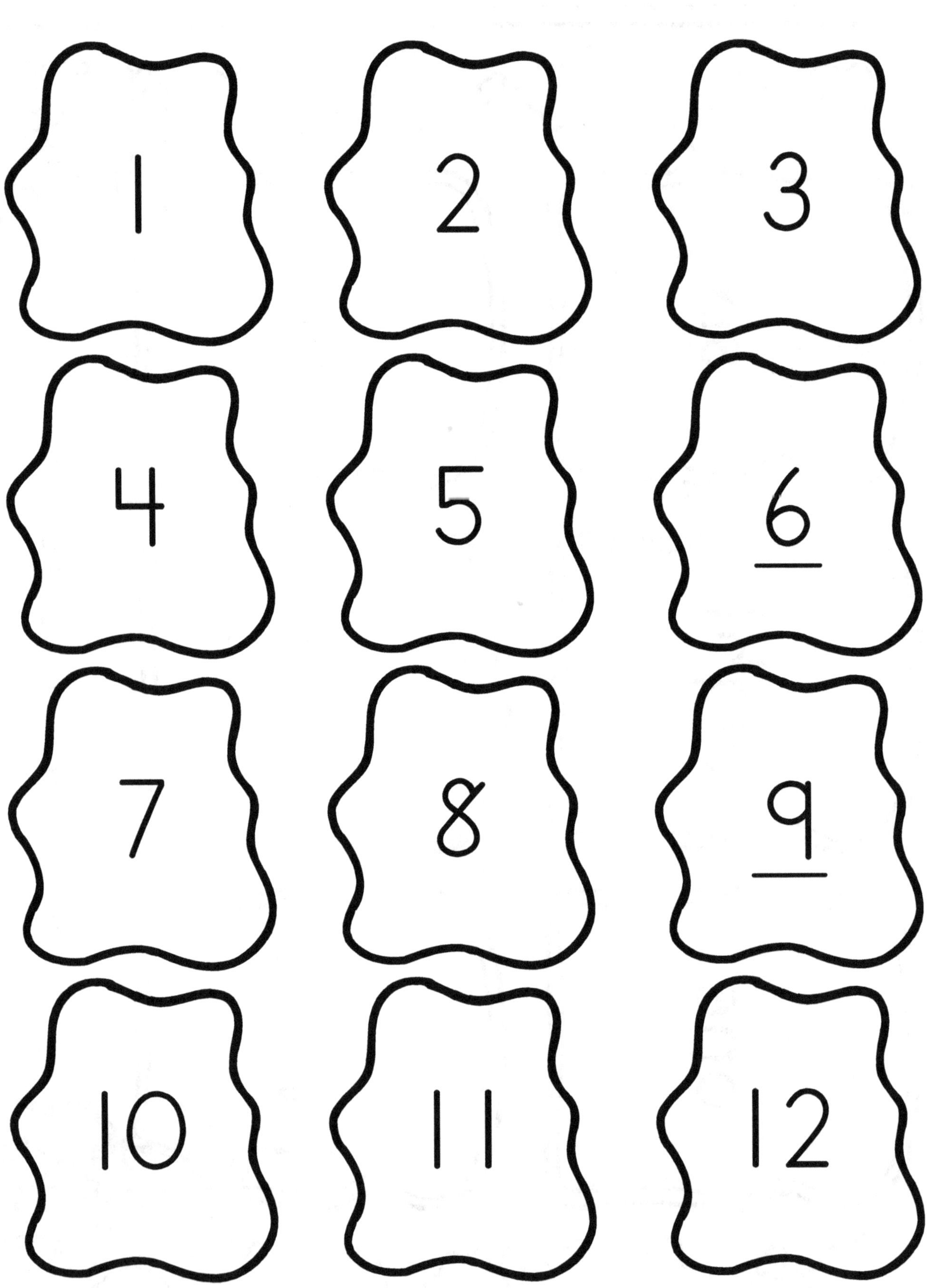

1	2	3
4	5	6
7	8	9
10	11	12

Reproduce,, color, and cut out a set of jelly game cards.

Jelly Game Cards

Reproduce,, color, and cut out a set of jelly game cards.

Polar Bear Patches
A Match Board Game
For Two Players

Materials
crayons, markers, scissors, glue, file folder, envelope

Assembly
Game Board: Reproduce, color, and cut out the cover and game board patterns. Assemble and glue the game board patterns to the inside of a folder. Glue the cover to the front of the folder, then laminate. Tape an envelope to the back of the game board folder to store game cards.

Game Cards: Reproduce, color, laminate, then cut out the game cards. Option: Reproduce, color, and glue a page of game cards to the back of a sheet of gift wrap, then laminate and cut out the cards. Store the game cards in the envelope on the back of the game board folder.

How to Play
Set up the game board on a table. Each player chooses a side of the board to play. One player shuffles and places the game cards, face down, on the table. Each player draws and places one card on his or her game board. Then each player, in turn, draws a card. If there is a match, the player identifies the match, and places the card on his or her polar bear. If there is no match, the player places the card, face down. When three matches are made, players take the matching patches off the game board and play continues until all the patches have been played. Reshuffle discarded patches if needed.

Polar Bear Patches Cover

Polar Bear Patches

Polar Bear Patches Game Board

Polar Bear Patches

Polar Bear Patches Game Board

Place three matching patches on the polar bear.

Patch Game Cards

Reproduce, color, and cut out one set of game cards.

50 LAB201311 • TEA FOR TWO • 978-1-937257-49-1 • © 2014 Little Acorn Books™

Patch Game Cards

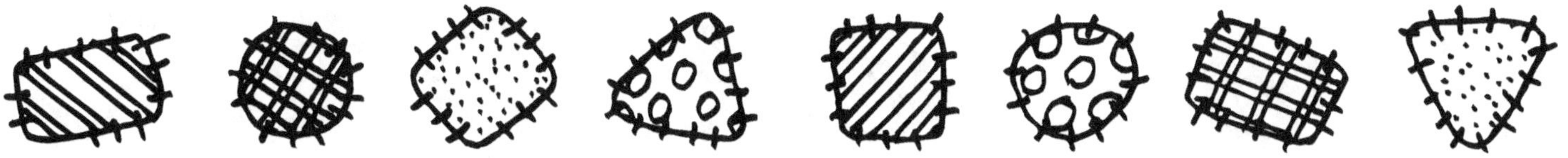

1	2	3
4	5	6
7	8	9
10	11	12

Creative Option: Provide children with construction paper doll cutouts.
Have children draw features on the cutouts, then color, cut out, and glue patches on the paper dolls.

Patch Game Cards

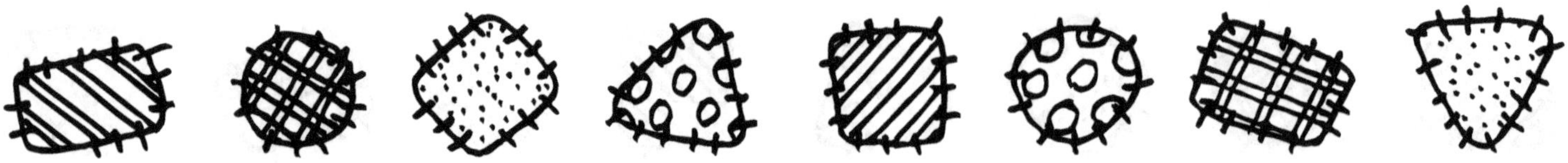

one two three
four five six
seven eight nine
ten eleven twelve

Reproduce, color, and cut out one set of game cards.

Cheep Cheep
A Match Board Game
For Two Players

Materials
crayons, markers, scissors, glue, file folder, envelope

Assembly
Game Board: Reproduce, color, and cut out the cover and game board patterns. Assemble and glue the game board patterns to the inside of a folder. Glue the cover to the front of the folder, then laminate. Tape an envelope to the back of the game board folder to store game cards.

Game Cards: Reproduce, color, laminate, then cut out the game cards. Option: Reproduce, color, and glue a page of game cards to the back of a sheet of gift wrap, then laminate and cut out the cards. Store the game cards in the envelope on the back of the game board folder. (Create non-matching game cards for advanced players, or all matches for early learners.)

How to Play
Set up the game board on a table. Players first choose to match either number word or number set chicks. One player shuffles and places the card deck, face down, on the table. Each player, in turn, draws a card. If there is a match, the player identifies the match, and places the card on the correct space in the basket. If there is no match, the player places the card, face down, in a discard pile. Play continues until there is a matching chick on each space in the basket.

Cheep Cheep Game Board

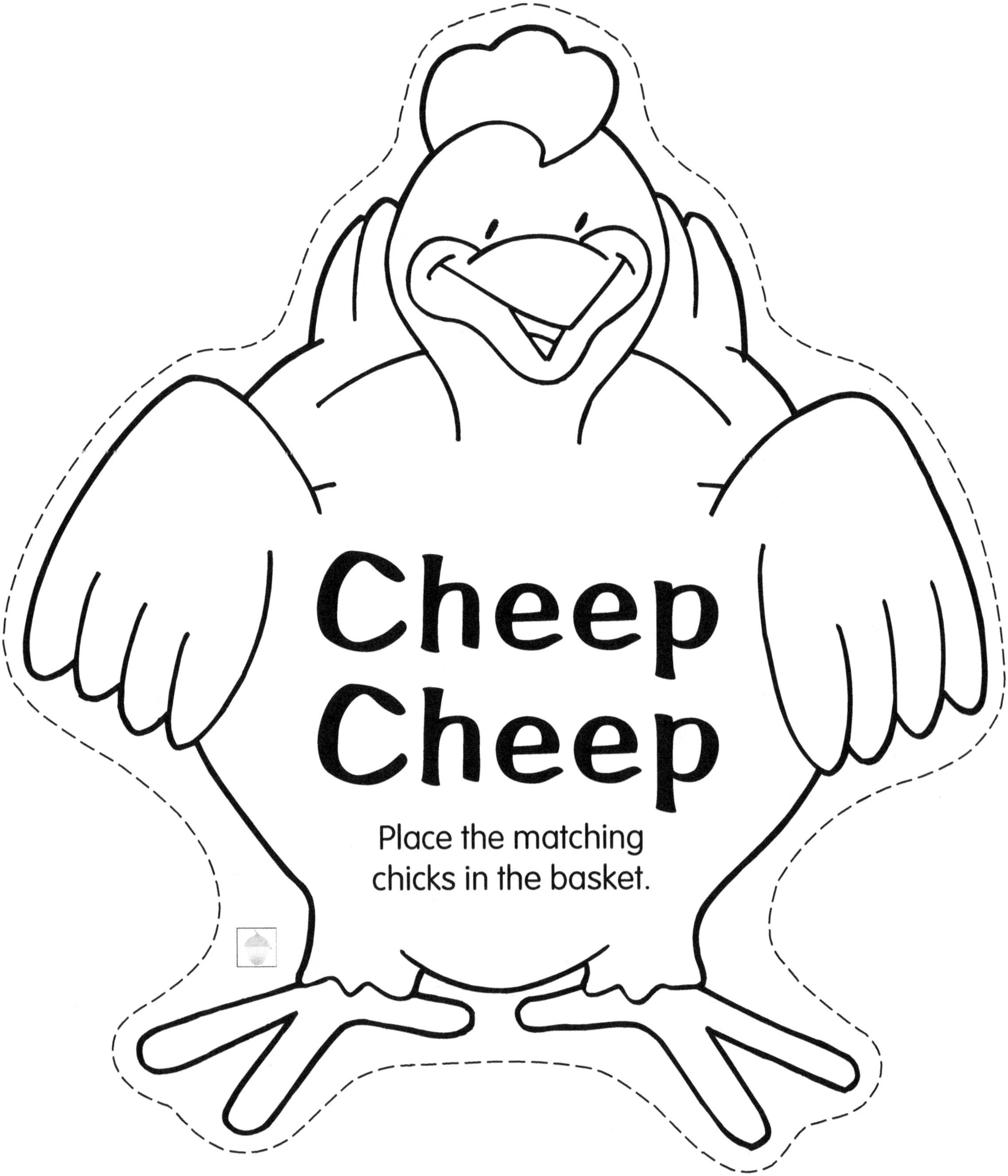

Cheep Cheep

Place the matching chicks in the basket.

Reproduce, color, cut out, and glue the hen to the top half of a file folder.

Cheep Cheep Game Board

Reproduce, color, cut out, and glue the basket to the bottom half of a file folder.

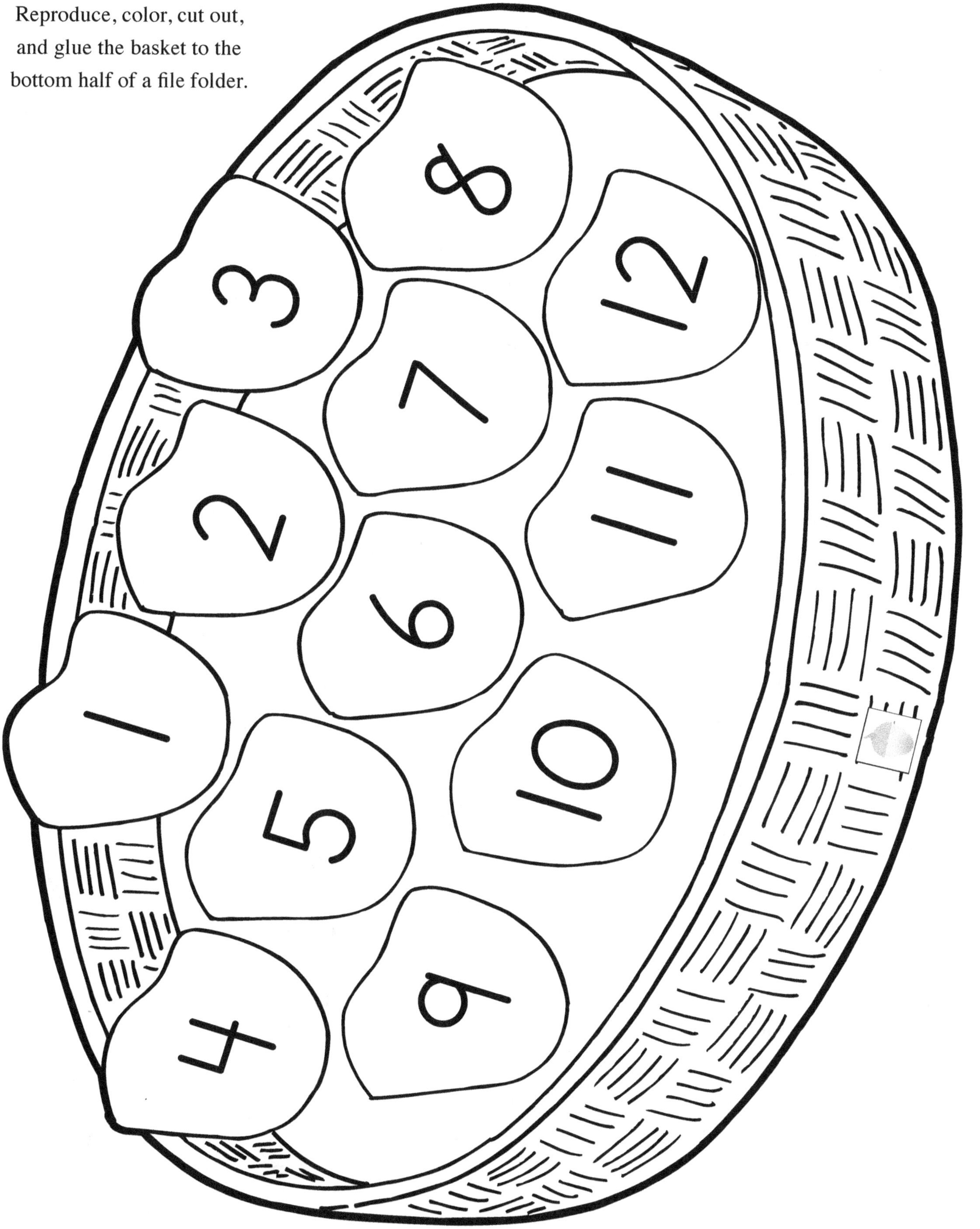

Chick Game Cards

Reproduce, color, cut out, and laminate.

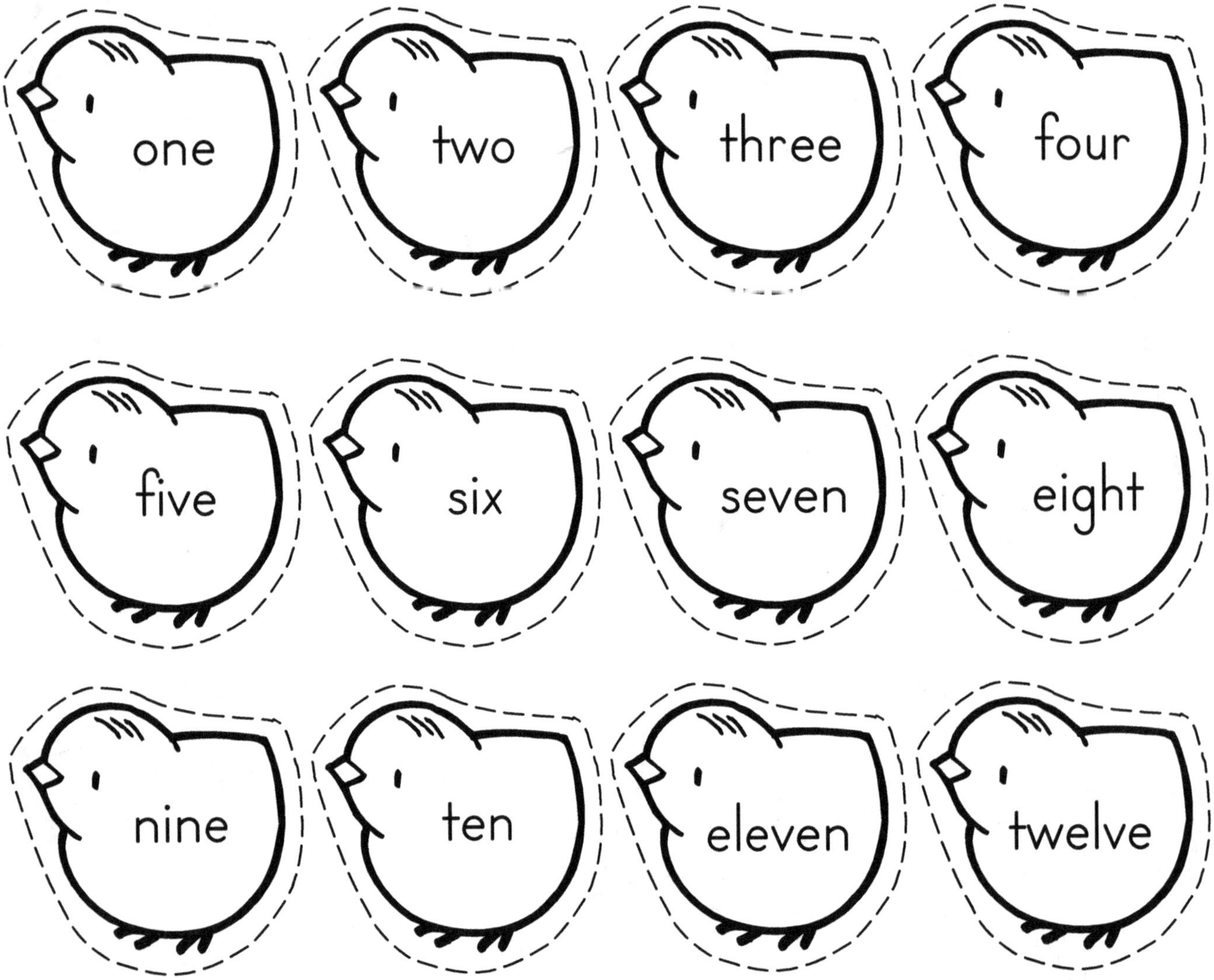

one	two	three	four
five	six	seven	eight
nine	ten	eleven	twelve

Creative Option: To make a name tag, reproduce, color, and cut out a chick. White out the number word and write in a child's name. Tape a safety pin to the back of the chick to form a pin.

Chick Game Cards

Reproduce, color, cut out, and laminate.

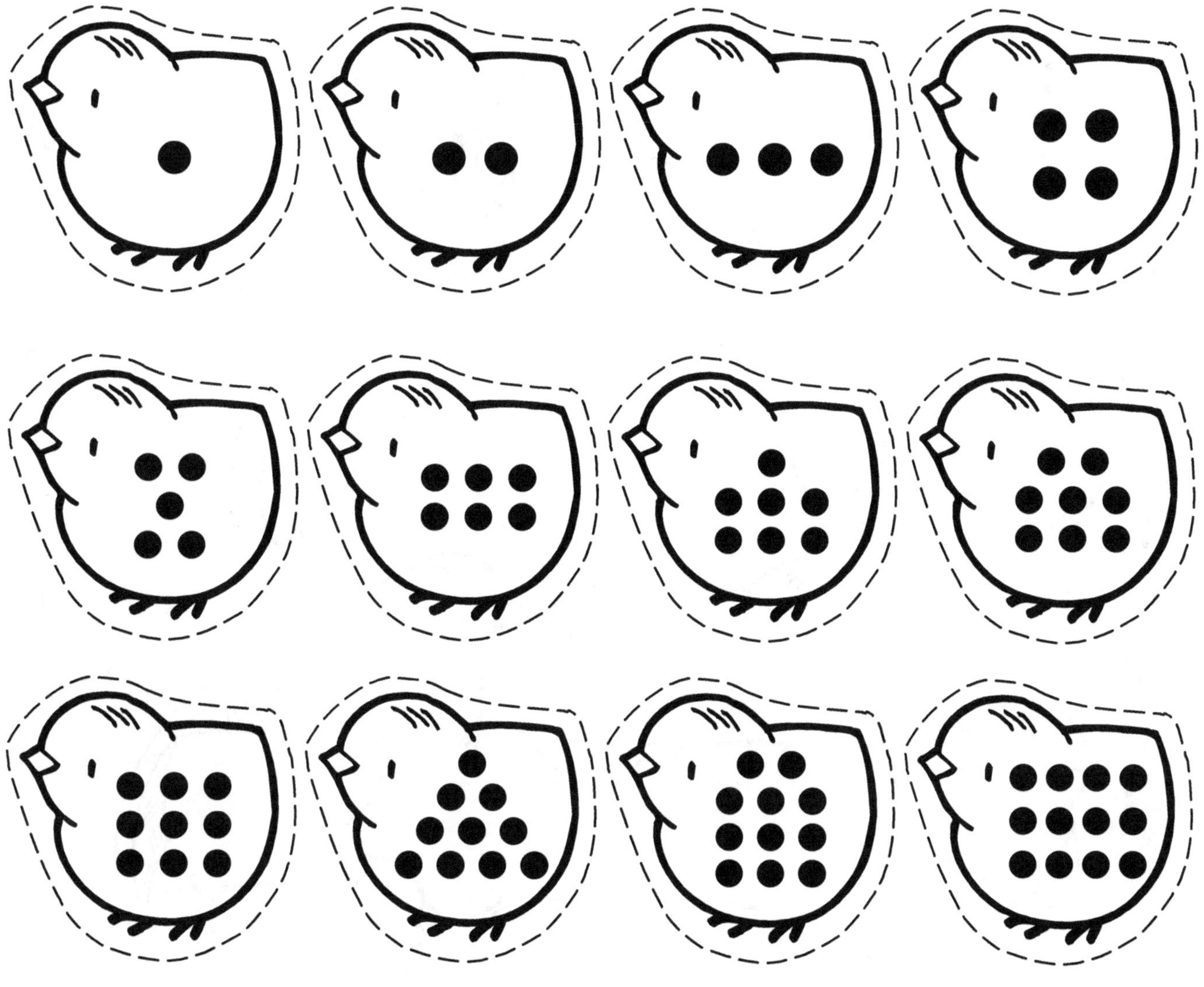

Creative Option: Photocopy two sets of cards to play a game of Concentration.

Cheep Cheep Cover

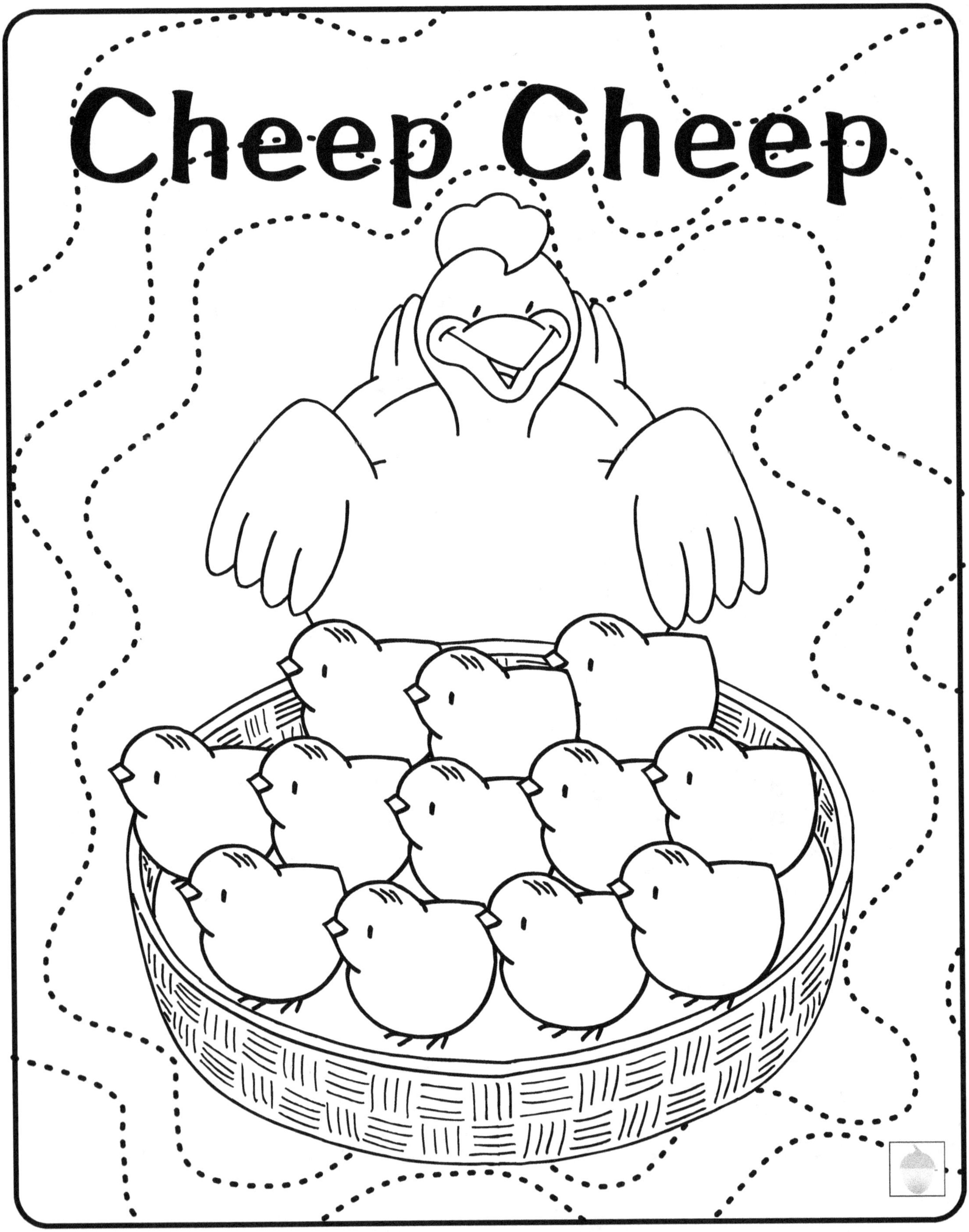

Cheep Cheep

Number Box Stackers
A Stacker Game
For One to Two Players

Materials
crayons, markers, scissors, glue, file folder, envelope

Assembly
Game Board: Reproduce, color, and cut out the cover and box stackers game board patterns. Glue each game board pattern to the inside of a folder. Glue the cover to the front of the folder, then laminate. Tape an envelope to the back of the game board folder to store game cards.

Game Cards: Reproduce, color, laminate, then cut apart two sets of game cards. Option: Reproduce, color, and glue each page of game cards to the back of a sheet of gift wrap, then laminate and cut apart the cards. Store the game cards in the envelope on the back of the game board folder.

How to Play
Set up the stacker game and cards on a table. Each player chooses a game board to play. Then one player shuffles and places the card deck, face down, on the table. Each player, in turn, draws, and stacks his or her drawn card on the matching space on the game board. Play continues until all the cards have been played.

Option: Use the game cards to play a game of Concentration. Shuffle and place all the cards, face down, on a table. Each player, in turn, turns over any two cards to find a match. If the player finds a match, he or she takes the cards and the next player takes a turn. If there is no match, each card is turned back over in the same position. Play continues until all the cards are taken.

Number Box Stackers Game Board

Number Box Stackers
Stack the matching boxes.

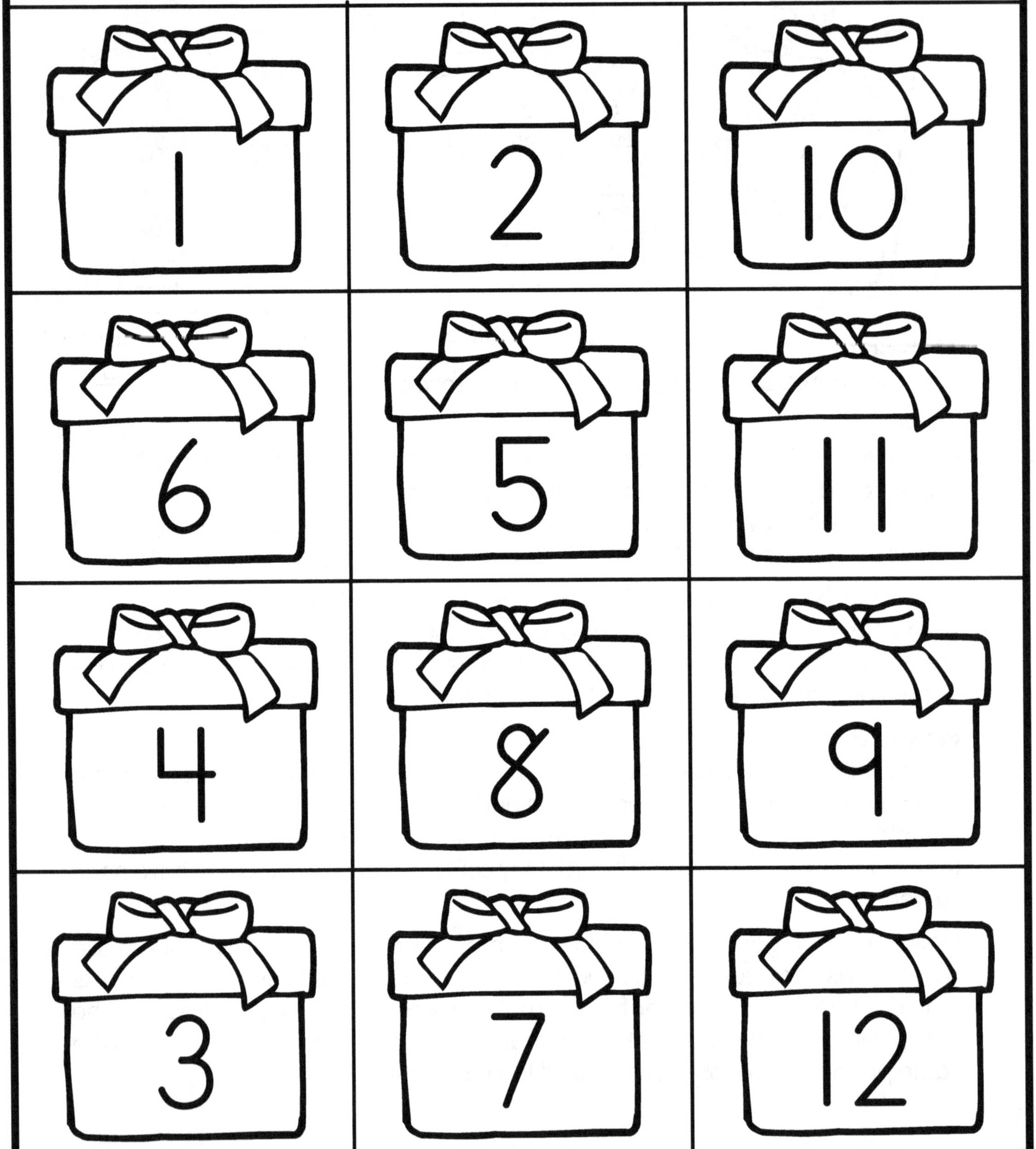

1	2	10
6	5	11
4	8	9
3	7	12

Number Box Stackers Game Board

Number Box Stackers
Stack the matching boxes.

9	8	6
2	10	11
4	1	5
3	7	12

Box Game Cards

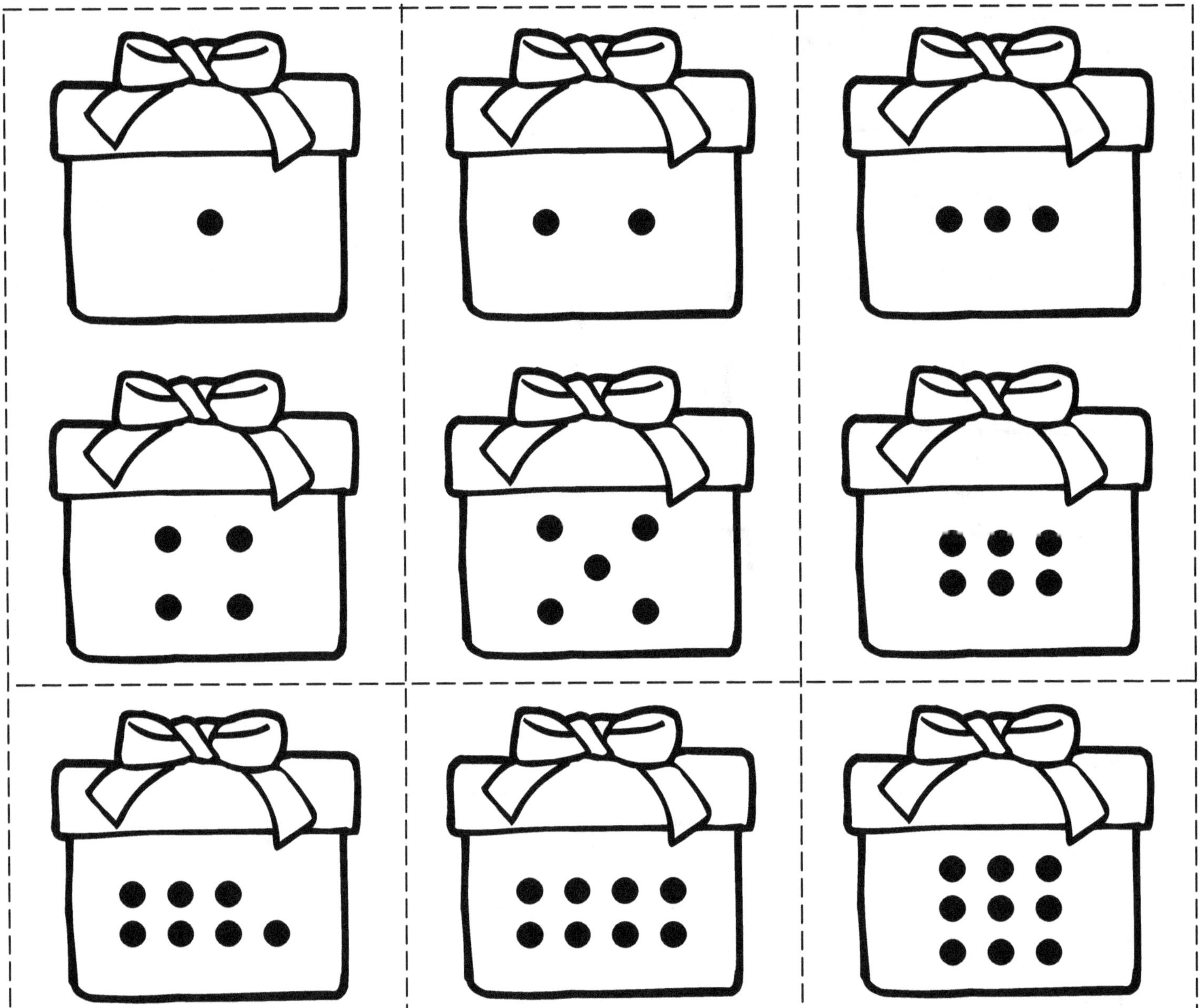

Reproduce two sets of game cards.

Box Game Cards

one	two	three
four	five	six
seven	eight	nine
ten	eleven	twelve

Reproduce two sets of game cards.

 Number Box Stackers Cover

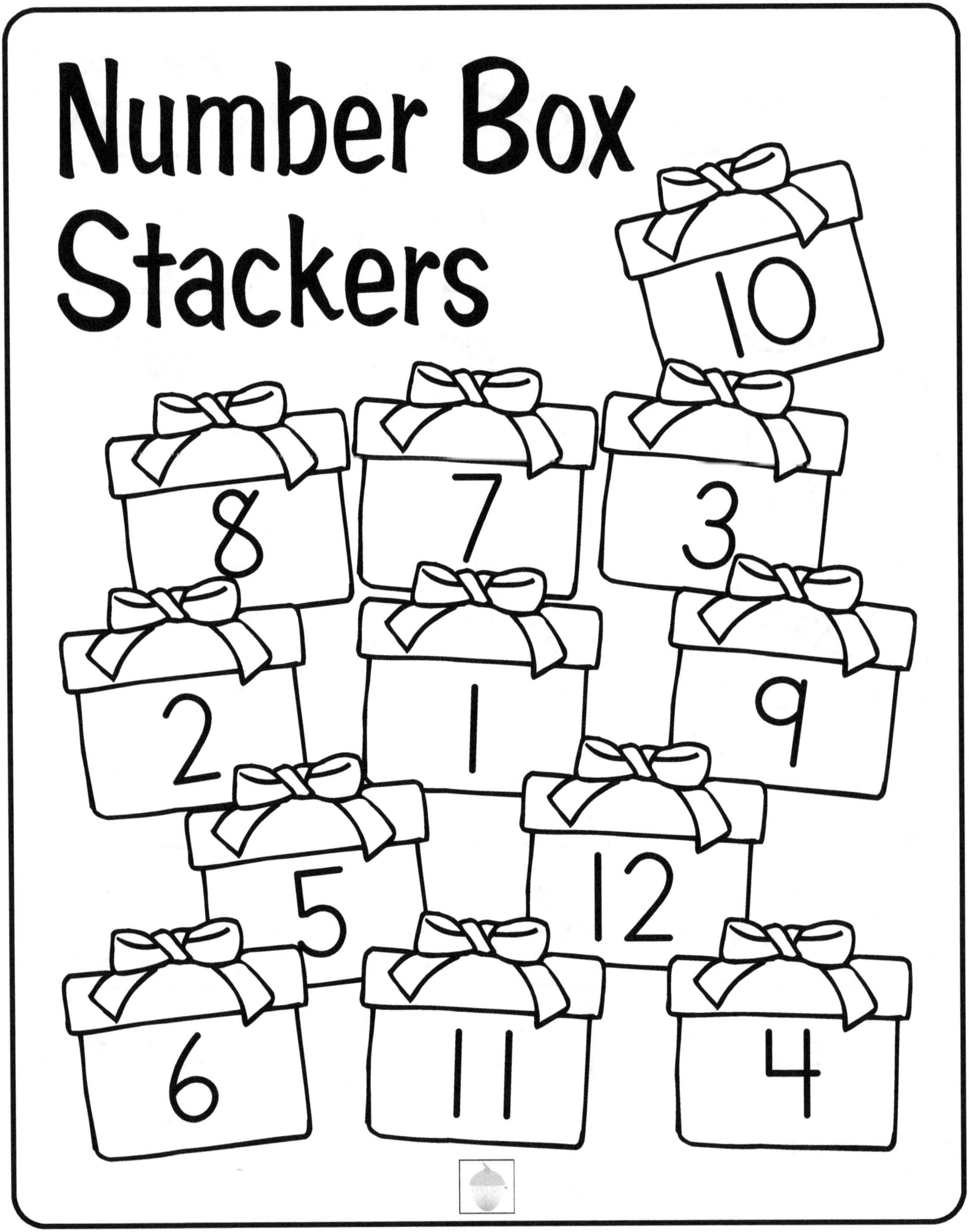

Number Box Stackers

Little Acorn Books™

Promoting Early Skills for a Lifetime™

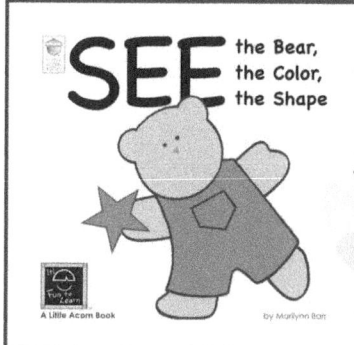   

A Hands-on Picture Book Series • Infancy–Age 4

Using Crayons, Scissors, & Glue for Crafts
Preschool–Grade 1

Miss Pitty Pat & Friends
Preschool–Grade 1

Mookie's Christmas Tree
For All Ages and Not Just for Christmas

Little Acorn Books™
Visit our web site:
www.littleacornbooks.com

LAB201311 • TEA FOR TWO • 978-1-937257-49-1 • © 2014 Little Acorn Books™

www.ingramcontent.com/pod-product-compliance
Lightning Source LLC
Chambersburg PA
CBHW081021040426

42444CB00014B/3300